Springer Theses

Recognizing Outstanding Ph.D. Research

Aims and Scope

The series "Springer Theses" brings together a selection of the very best Ph.D. theses from around the world and across the physical sciences. Nominated and endorsed by two recognized specialists, each published volume has been selected for its scientific excellence and the high impact of its contents for the pertinent field of research. For greater accessibility to nonspecialists, the published versions include an extended introduction, as well as a foreword by the student's supervisor explaining the special relevance of the work for the field. As a whole, the series will provide a valuable resource both for newcomers to the research fields described, and for other scientists seeking detailed background information on special questions. Finally, it provides an accredited documentation of the valuable contributions made by today's younger generation of scientists.

Theses are accepted into the series by invited nomination only and must fulfill all of the following criteria

- They must be written in good English.
- The topic should fall within the confines of Chemistry, Physics, Earth Sciences, Engineering and related interdisciplinary fields such as Materials, Nanoscience, Chemical Engineering, Complex Systems and Biophysics.
- The work reported in the thesis must represent a significant scientific advance.
- If the thesis includes previously published material, permission to reproduce this must be gained from the respective copyright holder.
- They must have been examined and passed during the 12 months prior to nomination.
- Each thesis should include a foreword by the supervisor outlining the significance of its content.
- The theses should have a clearly defined structure including an introduction accessible to scientists not expert in that particular field.

More information about this series at http://www.springer.com/series/8790

Zhijie Chen

Solutions of Nonlinear Schrödinger Systems

Doctoral Thesis accepted by
Tsinghua University, Beijing, China

 Springer

Author
Dr. Zhijie Chen
Department of Mathematical Sciences
Tsinghua University
Beijing
China

Supervisor
Prof. Wenming Zou
Department of Mathematical Sciences
Tsinghua University
Beijing
China

Present address
Dr. Zhijie Chen
Center for Advanced Study in Theoretical
  Sciences
National Taiwan University
Taipei
Taiwan

ISSN 2190-5053   ISSN 2190-5061   (electronic)
Springer Theses
ISBN 978-3-662-51542-6   ISBN 978-3-662-45478-7   (eBook)
DOI 10.1007/978-3-662-45478-7

Printed on acid-free paper

Springer-Verlag GmbH Berlin Heidelberg is part of Springer Science+Business Media
(www.springer.com)

Supervisor's Foreword

Although the real world seems in a muddle, many phenomena can be described by using nonlinear differential equations. For example, the incoherent solitons in nonlinear optics and the multispecies Bose–Einstein condensates (BEC) in hyperfine spin states, can be well characterized by some kinds of coupled nonlinear Schrödinger systems. By studying such systems, the propagation of self-trapped mutually incoherent wave packets and phase separation can be forecasted, though the physical experimental confirmation sometimes is posterior to the theoretical calculation for many years. The story of BEC is the best testimony.

For this reason, the nonlinear Schrödinger systems have received considerable attention in both physics and mathematics in the last decades. In particular, the research for such systems from the point of mathematics provides a theoretical enlightenment for those physical phenomena.

On the other hand, these nonlinear Schrödinger systems pose a lot of interesting but also challenging mathematical problems, which require people to develop new and deep theories and methods to treat them. For example, for the so-called BEC system, which has cubic nonlinearities and is weakly coupled, the least energy and the ground state have been attracting both physicists and mathematicians. With the deepening of the study on this line, some tough nuts remain uncracked.

Of course, there is also another different kind of nonlinear Schrödinger systems, the so-called linearly coupled systems. Systems of this type arise in nonlinear optics. For example, the propagation of optical pulses in nonlinear dual-core fiber can be described by two linearly coupled nonlinear Schrödinger equations. So far, there are still many important and hard problems on this topic remaining unsolved. In particular, the essential difference makes the study of these two kinds of systems quite different.

This thesis is mainly devoted to study several remaining important problems for nonlinear Schrödinger systems. For the BEC system in the subcritical case, qualitative properties of ground state solutions, including the uniqueness, asymptotic behaviors and the optimal parameter range for the existence are investigated. These results give the first partial answers to some open questions raised by Ambrosetti,

Colorado, and Sirakov. For the critical case, a systematical research on ground state solutions, including the existence, nonexistence, uniqueness, and phase separation phenomena of the limit profile is presented, which seem to be the first results for the BEC system in the critical case. Some delicate estimates on the least energy can be observed. It is well known that such equations become thorny due to the appearance of the critical exponent. The Palais-Smale compactness condition does not hold due to the lack of embedding compactness of the Sobolev space.

Moreover, some rather different phenomena are also discovered for more general critical systems of higher dimensional cases, which are proved to be closely related to the classical Brezis–Nirenberg critical exponent problem. For the linearly coupled system with critical exponent, an almost optimal result on the existence and nonexistence of the ground state solutions for different coupling constants is also proved.

All these results are mainly established via variational methods. Some new ideas and techniques are developed to treat the above problems. Some long-standing and puzzling questions are clarified in the current thesis. Simultaneously, it provides an excellent way to show the readers how the powerful variational tools can be applied to various elliptic systems.

Beijing, China, April 2014   Prof. Wenming Zou

Acknowledgments

First, I would like to express my sincere gratitude to my thesis advisor Professor Wenming Zou for his patient guidance, constant encouragement, and long-term help during my 5 years' graduate studies. He always inspired me to do challenging and also meaningful research. He is absolutely an honorable mathematician.

I would like to thank Professor Chang-Shou Lin for sharing his ideas and many valuable discussions. I also would like to thank my thesis committee members, Professor Shangquan Bu, Professor Yanheng Ding, Professor Yuxia Guo, Professor Shujie Li, and Professor Zhi-Qiang Wang, for kindly accepting to be the members of my thesis committee and helpful comments on my thesis.

I also would like to thank the Department of Mathematical Sciences, Tsinghua University, for its 9 years' education (both undergraduate and graduate) and support.

I also would like to thank my colleagues at Tsinghua University for their kind help and friendship.

Finally but not least, I would like to thank my parents, sister, and brother for their encouragement and support. My sincere thanks go to my beloved Erjuan Fu. Her unwavering love always accompany and encourage me.

Supplementary Note Several parts of this thesis were published in the following articles and reused with permission:

1. Z. Chen and W. Zou, Ground states for a system of Schrödinger equations with critical exponent, Journal of Functional Analysis. 2012; 262, 3091–3107. Copyright ©2012 Elsevier Masson SAS.
2. Z. Chen, C.-S. Lin and W. Zou, Multiple sign-changing and semi-nodal solutions for coupled Schrödinger equations, Journal of Differential Equations. 2013; 255, 4289–4311. Copyright ©2013 Elsevier Masson SAS.
3. Z. Chen and W. Zou, On the Brezis-Nirenberg problem in a ball. Differential and Integral Equations. 2012; 25, 527–542. Copyright ©2012 Khayyam Publishing Company.

Also some other parts of this thesis were published in several Springer mathematical journals (see the Reference). I have taken this opportunity to update some new materials (see the first section in every chapter) and references in this thesis to introduce some subsequent progress on several problems studied in this thesis after its completion.

Contents

Abbreviations

$\mathbb{R}^N$ The standard Euclidean space of dimension N

Ω A smooth bounded domain in $\mathbb{R}^N$

$\mathrm{supp}(u)$ The set $\overline{\{x \in \mathbb{R}^N | u(x) \neq 0\}}$

Δ The Laplace operator, namely $\Delta u = \frac{\partial^2 u}{\partial x_1^2} + \cdots + \frac{\partial^2 u}{\partial x_N^2}$

∇ The gradient operator, namely $\nabla u = \left(\frac{\partial u}{\partial x_1}, \cdots, \frac{\partial u}{\partial x_N} \right)$

2^* The Sobolev critical exponent, namely $2^* = \frac{2N}{(N-2)}$

$D^{1,2}(\mathbb{R}^N)$ Sobolev space $\left\{ u \in L^{2^*}(\mathbb{R}^N) : |\nabla u| \in L^2(\mathbb{R}^N) \right\}$

$H^1(\Omega)$ Sobolev space $\left\{ u \in L^2(\Omega) : |\nabla u| \in L^2(\Omega) \right\}$

$H^1_r(\mathbb{R}^N)$ $\left\{ u \in H^1(\mathbb{R}^N) : u \text{ is radially symmetric} \right\}$

$C_0^\infty(\Omega)$ $\left\{ u \in C^\infty(\Omega) : \mathrm{supp}(u) \text{ is a compact set of } \Omega \right\}$

$H_0^1(\Omega)$ The closure of $C_0^\infty(\Omega)$ with respect to the norm of $H^1(\Omega)$

Chapter 1
Introduction

Abstract In the last two decades, some kinds of coupled nonlinear Schrödinger systems have received a lot of attention from mathematicians, not only because they have great applications to various physical problems such as nonlinear optics and Bose–Einstein condensates, but also because they are very interesting and challenging in view of mathematics. This fact requires mathematicians to develop new ideas and new methods to investigate solutions of these nonlinear Schrödinger systems. So far, lots of mathematicians from all over the world have obtained many deep results for these nonlinear Schrödinger systems. However, there are still many important problems which remain open. In this thesis, we mainly study the existence and qualitative properties of nontrivial solutions for two kinds of nonlinear Schrödinger systems via variational methods and elliptic PDE theories. In this chapter, we give a basic introduction of the problems that we will study in the following chapters.

1.1 Mathematical Model Arising from Physics

The Schrödinger equation is the most important equation in quantum mechanics. The existence and qualitative properties of solutions to the Schödinger equation have always be one of the hottest research topics in both physics and mathematics in the last century. On the other hand, some kinds of coupled nonlinear Schrödinger systems have received ever-increasing interest in the last decades, and have become new focused topics in both physics and mathematics. For example, let us consider the following nonlinear Schrödinger system:

$$\begin{cases} -i\frac{\partial}{\partial t}\Phi_1 = \Delta\Phi_1 + \mu_1|\Phi_1|^2\Phi_1 + \beta|\Phi_2|^2\Phi_1, & x \in \Omega, \ t > 0, \\ -i\frac{\partial}{\partial t}\Phi_2 = \Delta\Phi_2 + \mu_2|\Phi_2|^2\Phi_2 + \beta|\Phi_1|^2\Phi_2, & x \in \Omega, \ t > 0, \\ \Phi_j = \Phi_j(x,t) \in \mathbb{C}, & j = 1, 2, \\ \Phi_j(x,t) = 0, & x \in \partial\Omega, \ t > 0, \ j = 1, 2, \end{cases} \quad (1.1)$$

where $\Omega \subset \mathbb{R}^N$ is a smooth bounded domain or $\Omega = \mathbb{R}^N$, i is the imaginary unit, $\mu_1, \mu_2 > 0$ and $\beta \neq 0$ are all parameters. This system is well known as coupled Gross–Pitaevskii equations in the literature (see [23, 76] for instance). System (1.1)

© Springer-Verlag Berlin Heidelberg 2015
Z. Chen, *Solutions of Nonlinear Schrödinger Systems*, Springer Theses,
DOI 10.1007/978-3-662-45478-7_1

has received a lot of attention from many physicists and mathematicians in the last 20 years, since it has great applications to various physical problems. First, system (1.1) appears as mathematical models in the research of incoherent solitons in nonlinear optics. For experimental results, we refer the reader to [70, 71] and references therein. Physically, the solution Φ_j denotes the jth component of the beam in Kerr-like photorefractive media [3]. The positive constant μ_j is for self-focusing in the jth component of the beam, and the coupling constant β is the interaction between the two components of the beam. System (1.1) also arises in the Hartree–Fock theory for a double condensate, i.e., a binary mixture of Bose–Einstein condensates in two different hyperfine states $|1\rangle$ and $|2\rangle$ [51]. Physically, Φ_j are the corresponding condensate amplitudes, μ_j and β are the intraspecies and interspecies scattering lengths. Precisely, the sign of μ_j represents the self-interactions of the single state $|j\rangle$. If $\mu_j > 0$ as considered here, it is called the focusing case, in opposition to the defocusing case where $\mu_j < 0$. Besides, the sign of β determines whether the interactions of states $|1\rangle$ and $|2\rangle$ are repulsive or attractive, i.e., the interaction is attractive if $\beta > 0$, and the interaction is repulsive if $\beta < 0$, where the two states are in strong competition when β is negative and very large [85]. See also [52, 59] and references therein for further details about the physical background of system (1.1).

In view of physics, solutions of $(\Phi_1(x, t), \Phi_2(x, t)) := (e^{i\lambda_1 t} u(x), e^{i\lambda_2 t} v(x))$ type with u, v both real-valued functions are called *solitary wave solutions*. They are surely very important in physics and have been widely investigated in the literature. By taking the expressions of solitary wave solutions into system (1.1), we obtain the following elliptic system:

$$\begin{cases} -\Delta u + \lambda_1 u = \mu_1 u^3 + \beta u v^2, & x \in \Omega, \\ -\Delta v + \lambda_2 v = \mu_2 v^3 + \beta v u^2, & x \in \Omega, \\ u|_{\partial\Omega} = v|_{\partial\Omega} = 0. \end{cases} \tag{1.2}$$

Here, when $\Omega = \mathbb{R}^N$, the Dirichlet boundary condition $u|_{\partial\Omega} = v|_{\partial\Omega} = 0$ means

$$u(x) \to 0 \text{ and } v(x) \to 0, \quad \text{as } |x| \to \infty,$$

and we always assume $\lambda_1, \lambda_2 > 0$ in this entire space case. When Ω is a smooth bounded domain, we let $\lambda_1(\Omega)$ be the first eigenvalue of $-\Delta$ in Ω with Dirichlet boundary condition, and always assume $\lambda_1, \lambda_2 > -\lambda_1(\Omega)$. Consequently, operators $-\Delta + \lambda_j$ are both positive definite for $j = 1, 2$.

In recent years, the existence and related properties of solutions to system (1.2) have been studied intensively in view of mathematics. In the following, we first recall some known results, and then give precise descriptions of the problems that we will study in this thesis.

1.2 Overview of Previous Research

Observe that system (1.2) is a weakly coupled system, namely it has solutions such that one component of the solution is null while the other one is not. This fact leads us to give the following definition.

Definition 1.1 We call that a solution (u, v) of (1.2) is a *nontrivial solution* if both $u \neq 0$ and $v \neq 0$, a solution (u, v) is a *semi-trivial solution* if $(u, v) \neq (0, 0)$ and (u, v) is type of $(u, 0)$ or $(0, v)$.

Observe that (1.2) has two kinds of semi-trivial solutions $(\omega_1, 0)$ and $(0, \omega_2)$, where ω_i are nontrivial solutions of the following scalar equation

$$- \Delta u + \lambda_i u = \mu_i u^3, \quad u > 0 \text{ in } \Omega, \quad u|_{\partial\Omega} = 0. \tag{1.3}$$

Clearly people are only concerned with the existence and properties of *nontrivial* solutions to (1.2). The existence of *semi-trivial* solutions makes the study of *nontrivial* solutions very challenging, which requires people to develop new ideas and approaches. In recent years, many mathematicians have obtained a lot of interesting results in this direction; see [6, 7, 13–15, 48, 49, 56, 61, 65, 67, 72, 80, 84, 86, 87] and references therein. See also [36, 57, 62, 63, 69, 77] for semiclassical states or singularly perturbed settings.

1.2.1 Ground State Solutions

Ground state solutions (also called least energy solutions in the literature) are the most important among all kinds of solutions, since they have very deep physical significance. First, we give the definition of ground state solutions to system (1.2). It is well known that solutions of system (1.2) correspond to the critical points of a C^2 functional $E_\beta : H = H_0^1(\Omega) \times H_0^1(\Omega) \to \mathbb{R}$ given by

$$E_\beta(u, v) = \frac{1}{2} \int_\Omega (|\nabla u|^2 + \lambda_1 u^2 + |\nabla v|^2 + \lambda_2 v^2) - \frac{1}{4} \int_\Omega (\mu_1 u^4 + 2\beta u^2 v^2 + \mu_2 v^4).$$

$$\tag{1.4}$$

Similarly as [61], we define the Nehari type manifold of system (1.2) as

$$\mathcal{N}_\beta := \left\{ (u, v) \in H \mid u \neq 0, v \neq 0, \int_\Omega (|\nabla u|^2 + \lambda_1 u^2) = \int_\Omega (\mu_1 u^4 + \beta u^2 v^2), \right.$$

$$\left. \int_\Omega (|\nabla v|^2 + \lambda_2 v^2) = \int_\Omega (\mu_2 v^4 + \beta u^2 v^2) \right\}. \tag{1.5}$$

Then any nontrivial solutions of (1.2) have to belong to $\mathcal{N}_\beta$. Take $\varphi, \psi \in C_0^\infty(\Omega)$ with $\varphi, \psi \not\equiv 0$ and $\text{supp}(\varphi) \cap \text{supp}(\psi) = \emptyset$, then there exist $t_1, t_2 > 0$ such that $(\sqrt{t_1}\varphi, \sqrt{t_2}\psi) \in \mathcal{N}_\beta$. So $\mathcal{N}_\beta \neq \emptyset$. Define the least energy

$$A_\beta := \inf_{(u,v)\in\mathcal{N}_\beta} E_\beta(u,v) = \inf_{(u,v)\in\mathcal{N}_\beta} \frac{1}{4}\int_\Omega (|\nabla u|^2 + \lambda_1 u^2 + |\nabla v|^2 + \lambda_2 v^2). \quad (1.6)$$

Since the operators $-\Delta + \lambda_j$ are positive definite for $j = 1, 2$, it follows easily from Sobolev inequalities that $A_\beta > 0$ for any $\beta \in \mathbb{R}$. Now we give the definition of ground state solutions as in [61].

Definition 1.2 We say that a solution (u, v) of (1.2) is *a ground state solution or a least energy solution*, if (u, v) is nontrivial and $E_\beta(u, v) = A_\beta$. We say that A_β is *attained*, if $A_\beta = E_\beta(u, v)$ for some $(u, v) \in \mathcal{N}_\beta$.

In this thesis, we will omit the subscript β for convenience when there is no confusion arising. Remark that, if (u, v) is a ground state solution of (1.2), then $E(u, v) \leq E(\varphi, \psi)$ holds for any other nontrivial solution (φ, ψ) of (1.2). On the other hand, if A is attained by some $(u, v) \in \mathcal{N}_\beta$, then (u, v) is a ground state solution provided $\beta < \sqrt{\mu_1\mu_2}$ (see Proposition A in Chap. 2). However, we have no idea whether (u, v) is a solution or not in the case $\beta \geq \sqrt{\mu_1\mu_2}$. That is, whether $\mathcal{N}_\beta$ is a *natural constraint* of E_β or not for $\beta \geq \sqrt{\mu_1\mu_2}$ is still unknown, which remains as an open question. This provides an evidence that systems are more difficult than scalar equations in general.

Now we recall some important results about ground state solutions. When $\Omega = \mathbb{R}^N (N = 2, 3)$ and $\lambda_1, \lambda_2 > 0$, the first mathematical result about system (1.2) was given by Lin and Wei [61] in 2005. Their main result is following: if $\beta < 0$, then the least energy A can not be attained, namely (1.2) has no ground state solutions; if $0 < \beta < \beta_0$ (here $\beta_0 \in (0, \sqrt{\mu_1\mu_2})$ is a small constant), then the least energy A is attained, namely (1.2) has ground state solutions. In fact, they studied a more general system of m ($m \geq 2$) coupled equations; see [61] for details.

Later in 2007, Sirakov [80] made a deep research on the existence of ground state solutions to system (1.2) in the entire space case $\Omega = \mathbb{R}^N(N = 2, 3)$. In the symmetric case $\lambda_1 = \lambda_2$, he proved that: if $0 < \beta < \min\{\mu_1, \mu_2\}$ or $\beta > \max\{\mu_1, \mu_2\}$, then (1.2) has a ground state solution of type $(\sqrt{k}\omega, \sqrt{l}\omega)$, where (k, l) satisfies $\mu_1 k + \beta l = 1$ and $\mu_2 l + \beta k = 1$, and ω is a positive solution of (1.3) with $\mu_i = 1$ (which is unique up to a translation; see [60]); if $\min\{\mu_1, \mu_2\} \leq \beta \leq \max\{\mu_1, \mu_2\}$ and $\mu_1 \neq \mu_2$, then (1.2) has no nontrivial positive solutions; if $\mu_1 = \mu_2 = \beta$, then (1.2) has infinitely many positive solutions (u, v) with $u/v \equiv$ constant. Clearly, this is the optimal result about the existence of ground state solutions in the special case $\lambda_1 = \lambda_2$. In the general case $\lambda_1 \neq \lambda_2$, he proved that: there exist two positive constants $\beta_1 < \beta_2$ such that (1.2) has a ground state solution whenever $0 < \beta < \beta_1$ or $\beta > \beta_2$. Moreover, he gave the accurate expressions of β_1 and β_2. Finally, Sirakov asked an open question in [80]: What are the optimal ranges of the parameter β for the existence of ground state solutions in the general case $\lambda_1 \neq \lambda_2$? Remark that, [6, 7, 49] also contained similar existence results as those of [80] in the case $\lambda_1 \neq \lambda_2$ independently.

1.2.2 Multiple Solutions

Does system (1.2) admit any other solutions besides ground state solutions? This is a basic question about the existence of multiple solutions that has received great interest in the last 10 years. So far, there have been also many interesting results in this direction in the literature. First, we assume that $\Omega \subset \mathbb{R}^N (N = 2, 3)$ is a smooth bounded domain. In the case where $\lambda_1 = \lambda_2 > 0$, $\mu_1 = \mu_2 > 0$ and $\beta \leq -\mu_1$, Dancer et al. [48] proved the existence of infinitely many positive solutions for system (1.2), and the same result was proved later for the more difficult case $\lambda_1 = \lambda_2 < 0$ (note that the operators $-\Delta + \lambda_j$ are no longer positive definite) by Noris and Ramos [72]. When Ω is a ball, Wei and Weth [87] obtained an interesting result on positive radially symmetric solutions (the difference between two components of a solution has prescribed zero numbers). Remark that, since $\lambda_1 = \lambda_2$ and $\mu_1 = \mu_2$, system (1.2) is invariant under the transformation $(u, v) \mapsto (v, u)$, which plays a crucial role in the proof of [48, 72, 87]. Later, by using Rabinowitz's global bifurcation approach, the result of [87] was reproved by Bartsch et al. [13] without requiring the symmetric condition $\mu_1 = \mu_2$, but in their proof the assumption $\lambda_1 = \lambda_2$ plays a crucial role. Recently, Guo and Wei [55] reproved the result of [48] in the case $N = 2$ and $\mu_1 = \mu_2$ but without assuming $\lambda_1 = \lambda_2$. On the other hand, under assumptions $\lambda_i > 0$, $\mu_i > 0$ and $\beta < 0$ without requiring either $\lambda_1 = \lambda_2$ or $\mu_1 = \mu_2$, Sato and Wang [78] proved that system (1.2) has infinitely many semi-positive solutions (namely at least one component of a solution is positive, but whether the other one positive or sign-changing is unknown). Now we consider the entire space case $\Omega = \mathbb{R}^N$. Under assumptions $\lambda_i > 0$, $\mu_i > 0$ and $\beta < 0$, Liu and Wang [65] proved that system (1.2) has infinitely many nontrivial solutions. Remark that whether solutions obtained in [65] are positive or sign-changing are not known. When $\beta > 0$ is sufficiently large, radially symmetric sign-changing solutions of (1.2) with prescribed zeros were obtained in [68]. All the results mentioned above can be seen in [13, 48, 55, 65, 68, 78, 84, 86, 87].

Besides, qualitative properties of solutions have also been well investigated in the literature. For example, the uniqueness of positive solutions to (1.2) under some special circumstances was studied in [89], while [46, 48, 73, 74, 84, 87, 88] studied other properties of solutions to (1.2), such as a priori estimates of positive solutions, the asymptotic behavior of solutions as $\beta \to -\infty$, the regularity of the limit solutions, phase separation phenomena of the limiting profile, and so on.

1.3 Problems Studied in This Thesis

As pointed out above, mathematicians have obtained many interesting results about system (1.2) in the last decades. However, there are still many open problems about (1.2), which are of great interest to many researchers. In this thesis, we mainly study several important problems which are closely related to either system (1.2) or a linearly coupled Schrödinger system.

First, as far as we know, there seems *no* any answers to the open question raised by Sirakov [80]. Therefore, the first problem we plan to study is:

1. Under some special circumstances, we investigate the optimal parameter range of β for the existence of ground state solutions to system (1.2) in the general case $\lambda_1 \neq \lambda_2$. This is surely a very interesting problem and seems also very difficult. We will give the first partial answer in Chap. 2. Moreover, among other things, we also prove the uniqueness of ground state solutions in the symmetric case $\lambda_1 = \lambda_2$. We believe that our results will be important complements to the study of ground state solutions to system (1.2).

On the other hand, we know from the above introduction that, there have been many deep results on positive solutions of (1.2), such as properties of ground state solutions and multiplicity of positive solutions. However, there are very few results about sign-changing solutions in the literature, since the study of sign-changing solutions seems to be more tough; see Chap. 3 for further details. Here, the second problem we plan to study is:

2. When $\beta > 0$ is small, we prove the existence of multiple sign-changing solutions and semi-nodal solutions (namely one component of a solution changing sign and the other one positive) to system (1.2); see Chap. 3.

Now, let us define the Sobolev critical exponent $2^* = \frac{2N}{N-2}$ for $N \geq 3$. Remark that, all references mentioned above deal with the case $N \leq 3$. In this case, the nonlinearities and coupling terms of (1.2) are all subcritical, and so the well-known Palais–Smale condition of the corresponding functional holds (for the entire space case, it suffices to restrict the functional on the Sobolev subspace of radially symmetric functions), which makes the study of system (1.2) via variational methods slightly easy. On the other hand, in 1983 Brezis and Nirenberg [19] studied the following critical exponent problem:

$$- \Delta u + \lambda u = u^{2^*-1}, \quad u > 0 \text{ in } \Omega, \quad u|_{\partial\Omega} = 0, \tag{1.7}$$

where Ω is a smooth bounded domain in $\mathbb{R}^N$ with $N \geq 3$. This is the well-known Brezis–Nirenberg problem. From then on, various critical exponent problems, including scalar equations and coupled systems, have received ever-increasing interest and have been widely studied in the past 30 years. One reason is that, critical exponent problems, which lack the compactness property required for the validity of the Palais–Smale condition, are very interesting and challenging in view of mathematics. Comparing to subcritical problems, there are usually some quite different phenomenon appearing when studying critical exponent problems. Another reason is that, critical exponent problems are closely related to problems arising from conformal geometry, such as the well-known Yamabe problem and conformal scalar curvature equations. Thus, critical exponent problems are surely worthy to be studied intensively. Recently, coupled nonlinear Schrödinger systems with critical exponents and Hardy potentials in the entire space $\mathbb{R}^N$ were well studied in [1, 39, 41]. However, to the best of our knowledge, there seems *no* any result about system (1.2) in the critical case $N = 4$. Remark that when $N = 4$, if $\Omega = \mathbb{R}^N$ and $\lambda_1 \lambda_2 > 0$, then it easily follows from the Pohozaev identity that (1.2) has no nontrivial solutions. Therefore, when $N = 4$, we only consider the case where $\Omega \subset \mathbb{R}^4$ is a smooth bounded domain. The third problem we plan to study is:

3. In the critical case $N = 4$, we make a systematical study on the ground state solutions of (1.2), including existence, nonexistence, uniqueness, and asymptotic behaviors. See Chap. 4. Our study will be the first contribution to system (1.2) in the critical case $N = 4$.

In general, the research of critical exponent problems depends heavily on the space dimension N, and the answers of the same question might be quite *different* for different spatial dimensions, particularly between low dimensions and high dimensions. For example, for the Brezis–Nirenberg problem (1.7), Brezis and Nirenberg [19] proved that, for $N \geq 4$ and $-\lambda_1(\Omega) < \lambda < 0$, (1.7) has a positive solution; but for the case where Ω is the unit ball in $\mathbb{R}^3$ and $-\lambda_1(\Omega)/4 \leq \lambda < 0$, (1.7) has no positive solutions. Motivated by this fact, there are some natural questions we may ask: Can we extend some results in dimension four case to a homologous critical system with spatial dimensions $N \geq 5$? Will the conclusions be different between these two cases? Are there any differences in the research methods between these two cases? These are surely very interesting questions in view of mathematics. Thus, the fourth problem we plan to study is:

4. In the case $N \geq 5$, we study ground state solutions of the following homologous critical system of (1.2):

$$\begin{cases} -\Delta u + \lambda_1 u = \mu_1 |u|^{2^*-2} u + \beta |u|^{\frac{2^*}{2}-2} u |v|^{\frac{2^*}{2}}, & x \in \Omega, \\ -\Delta v + \lambda_2 v = \mu_2 |v|^{2^*-2} v + \beta |v|^{\frac{2^*}{2}-2} v |u|^{\frac{2^*}{2}}, & x \in \Omega, \\ u|_{\partial\Omega} = v|_{\partial\Omega} = 0. \end{cases} \tag{1.8}$$

Observe that when $\beta = 0$, system (1.8) is just the Brezis–Nirenberg problem (1.7). As pointed out before, the Brezis–Nirenberg problem (1.7) depends heavily on spatial dimensions (see also [30, 32, 50, 79] for more evidences). Clearly, system (1.8) can be seen as a coupled perturbed Brezis–Nirenberg problem. Thus, the existence of positive solutions of system (1.8) should also depend on the spatial dimensions. More interestingly, positive solutions of system (1.8) might be related to sign-changing solutions of the Brezis–Nirenberg problem (1.7). Therefore, there should be some differences between $N = 4$ and $N \geq 5$ for system (1.8). We will investigate these questions in Chap. 5. Besides, we will also give sharp estimates of the least energy to the Brezis–Nirenberg problem (1.7) in open balls, which will be used in the proof of asymptotic behaviors of the ground state solutions to system (1.8). Of course, the result about sharp energy estimates for the Brezis–Nirenberg problem is interesting in itself, and will be an important complement to the study of such a classical problem.

Finally, observe that the coupling terms in systems (1.2) and (1.8) are all of nonlinear growth. Of course, there are also many interesting elliptic systems which are coupled in other ways, such as the well-known Lane–Emden system. In 2006, Ambrosetti and Colorado [6] introduced the following linearly coupled system:

$$
\begin{cases}
-\Delta u + \mu u = |u|^{p-1}u + \lambda v, & x \in \mathbb{R}^N, \\
-\Delta v + \nu v = |v|^{q-1}v + \lambda u, & x \in \mathbb{R}^N, \\
u, v \to 0 \text{ as } |x| \to +\infty,
\end{cases}
\tag{1.9}
$$

where μ, ν, λ are positive constants and $p, q > 1$, and $p, q \le 2^* - 1$ if $N \ge 3$. This system also arises in nonlinear optics [4]. A basic difference between (1.9) and (1.2) [also (1.8)] is that (1.9) has no semi-trivial solutions [namely $(u, 0)$ or $(0, v)$] because of the linearly coupling terms, that is, (1.9) is a fully coupled system but (1.2) and (1.8) are weakly coupled systems. In the subcritical case $1 < p, q < 2^* - 1$, Ambrosetti and his coauthors obtained several interesting results for this system; see [5, 8, 9]. More details will be given in Chap. 6. To the best of our acknowledge, there seems *no* any result for system (1.9) in the critical case. Remark that, if $p = q = 2^* - 1$, then we easily deduce from the Pohozaev identity that (1.9) has no nontrivial solutions. Hence, the fifth problem we plan to study is:

5. Under assumptions $1 < p < q = 2^* - 1$, we study the ground state solutions of system (1.9), such as existence, nonexistence and asymptotic behaviors. Moreover, our result on the existence and nonexistence is almost *optimal*. See Chap. 6.

We will study the above five problems via variational methods and elliptic PDE theories. Since the birth of the calculus of variations, it has been realized that variational methods are very powerful and can be applied in a very large number of problems. One purpose of this thesis is to show the reader how variational methods might be applied in various elliptic systems. Remark that, for different problems, different difficulties might appear in the application of variational methods. We will develop some new ideas and approaches (such as the method of proving the uniqueness of ground state solutions in Chaps. 4 and 5) to deal with these problems. We believe that our new ideas will have applications in other problems.

In the rest of this thesis, we always denote positive constants (possibly different in different places) by $C, C_0, C_1, \dots$.

Chapter 2
A BEC System with Dimensions $N = 2, 3$: Ground State Solutions

Abstract As introduced in Chap. 1, we study the ground state solutions of system (1.2) in the entire space $\mathbb{R}^N$ with $N = 2, 3$. Precisely, motivated by Sirakov's previous work, we prove some uniqueness results of positive (ground state) solutions for the special case $\lambda_1 = \lambda_2$. These give partial answers to Sirakov's conjecture. For the general case $\lambda_1 \neq \lambda_2$, we prove a sharp result on the parameter range for the existence of ground state solutions. The asymptotic behaviors of ground state solutions can be investigated as a corollary. We also prove a nonexistence result about positive solutions. These results answer partially some open questions raised by Ambrosetti, Colorado and Sirakov. Our proof is mainly applying asymptotic analysis together with the classical bifurcation theory.

2.1 Main Results

Consider the following system with cubic nonlinearities which arises as mathematical models from nonlinear optics and Bose-Einstein condensates (BEC):

$$\begin{cases} -\Delta u + \lambda_1 u = \mu_1 u^3 + \beta u v^2, & x \in \mathbb{R}^N, \\ -\Delta v + \lambda_2 v = \mu_2 v^3 + \beta v u^2, & x \in \mathbb{R}^N, \\ u \geq 0, v \geq 0 \text{ in } \mathbb{R}^N, \\ u, v \to 0 \text{ as } |x| \to +\infty, \end{cases} \tag{2.1}$$

where $N = 2, 3$, $\mu_1, \mu_2, \lambda_1, \lambda_2 > 0$, and $\beta \neq 0$ is a coupling constant. As pointed out in Chap. 1, system (2.1) has two semi-trivial solutions $(\omega_1, 0)$ and $(0, \omega_2)$, where ω_i is the unique positive radially symmetric solution of the following scalar equation (see [60])

$$-\Delta u + \lambda_i u = \mu_i u^3, \quad u > 0, \ u \in H^1(\mathbb{R}^N), \tag{2.2}$$

© Springer-Verlag Berlin Heidelberg 2015
Z. Chen, *Solutions of Nonlinear Schrödinger Systems*, Springer Theses,
DOI 10.1007/978-3-662-45478-7_2

and the corresponding least energy is

$$B_i := \frac{1}{2} \int_{\mathbb{R}^N} (|\nabla \omega_i|^2 + \lambda_i \omega_i^2) \, dx - \frac{1}{4} \int_{\mathbb{R}^N} \mu_i \omega_i^4 \, dx. \tag{2.3}$$

Clearly there holds

$$\int_{\mathbb{R}^N} (|\nabla u|^2 + \lambda_i u^2) \, dx \geq 2\sqrt{B_i} \left(\int_{\mathbb{R}^N} \mu_i u^4 \, dx \right)^{1/2}, \quad \forall u \in H^1(\mathbb{R}^N). \tag{2.4}$$

Denote $H := H^1(\mathbb{R}^N) \times H^1(\mathbb{R}^N)$ for convenience. Recalling that $E_\beta, \mathcal{N}_\beta, A_\beta$ are defined in (1.4), (1.5) and (1.6) respectively, and the ground state solution is defined in Definition 1.2. Sirakov [80] proved the following interesting result in 2007.

Theorem A ([80, Theorem 1]) *Suppose $\lambda_1 = \lambda_2 = \lambda$.*

(i) *For $\beta \in (0, \min\{\mu_1, \mu_2\}) \cup (\max\{\mu_1, \mu_2\}, +\infty)$, A_β is attained by the couple $(\sqrt{k_\beta}\omega_0, \sqrt{l_\beta}\omega_0)$, where (k_β, l_β) satisfies $\mu_1 k + \beta l = 1$ and $\mu_2 l + \beta k = 1$, and ω_0 is the unique positive radially symmetric solution of (2.2) with $\mu_i = 1$. That is, $(\sqrt{k_\beta}\omega_0, \sqrt{l_\beta}\omega_0)$ is a ground state solution of (2.1).*

(ii) *For $\beta \in [\min\{\mu_1, \mu_2\}, \max\{\mu_1, \mu_2\}]$ and $\mu_1 \neq \mu_2$, (2.1) has no nontrivial nonnegative solutions.*

Then Sirakov [80] raised a conjecture: For spatial dimensions $N = 1, 2, 3$, the couple $(\sqrt{k_\beta}\omega_0, \sqrt{l_\beta}\omega_0)$ is the unique positive solution to (2.1) up to a translation. Recently, Wei and Yao [89, Theorem 4.2] proved this conjecture in case $\beta > \max\{\mu_1, \mu_2\}$, and [89, Theorem 1.1] proved it in the case where $0 < \beta < \min\{\mu_1, \mu_2\}$ and $N = 1$. For $N = 2, 3$, this conjecture was also proved in [89, Theorem 4.1] for $0 < \beta < \beta'$, where β' is a small constant. That is, whether Sirakov's conjecture holds or not in the remaining case where $N = 2, 3$ and $\beta \in (\beta', \min\{\mu_1, \mu_2\})$ remains open. Here we can give a partial answer.

Theorem 2.1 *Assume that $\lambda_1 = \lambda_2$ and $\mu_1 \neq \mu_2$. Then there exists small $\delta > 0$ such that for any $\beta \in (\min\{\mu_1, \mu_2\} - \delta, \min\{\mu_1, \mu_2\})$, the couple $(\sqrt{k_\beta}\omega_0, \sqrt{l_\beta}\omega_0)$ is the unique positive solution of system (2.1) up to a translation.*

The proof of Theorem 2.1 mainly use the classical bifurcation theory. Clearly, neither [89, Theorem 4.1] nor Theorem 2.1 proves Sirakov's conjecture for all $\beta \in (0, \min\{\mu_1, \mu_2\})$. A slightly weaker but natural question is: for $\beta \in (0, \min\{\mu_1, \mu_2\})$, whether the *ground state* solutions are unique or not, up to a translation? Here we can give a positive answer to this question.

Theorem 2.2 *Assume $\lambda_1 = \lambda_2$. Then for any $\beta \in (0, \min\{\mu_1, \mu_2\})$, the couple $(\sqrt{k_\beta}\omega_0, \sqrt{l_\beta}\omega_0)$ is the unique ground state solution of system (2.1) up to a translation.*

Clearly there is a fully symmetric case where $\lambda_1 = \lambda_2$ and $\mu_1 = \mu_2 = \beta$ remaining. In this case, it is easy to check that $((2\beta)^{-1/2} \cos\theta\omega_0, (2\beta)^{-1/2} \sin\theta\omega_0)$ is a positive solution of (2.1) for any $\theta \in (0, \frac{\pi}{2})$. In fact, Wei and Yao [89, Theorem 1.2] proved that

$$\mathfrak{S} := \left\{ ((2\beta)^{-1/2} \cos\theta\omega_0, (2\beta)^{-1/2} \sin\theta\omega_0) : \theta \in (0, \frac{\pi}{2}) \right\} \tag{2.5}$$

contains all positive solutions of (2.1) for $N = 1$. Here we can prove a stronger result for $N = 2, 3$.

Theorem 2.3 *Let $N = 2, 3$, $\lambda_1 = \lambda_2$ and $\mu_1 = \mu_2 = \beta$. Assume that (u, v) be a nontrivial solution of (2.1) with $u > 0$. Then $v = Cu$ for some constant $C \neq 0$. In particular, the set $\mathfrak{S}$ contains all positive solutions of (2.1), and (2.1) has no semi-nodal solutions (namely one component of the solution positive and the other one sign-changing; the definition will be given Chap. 3).*

Now let us consider the general case $\lambda_1 \neq \lambda_2$. Clearly in this case, system (2.1) has *no* nontrivial solutions (u, v) satisfying $u/v \equiv$ constant. This fact makes the general case $\lambda_1 \neq \lambda_2$ much more delicate comparing to the symmetric case $\lambda_1 = \lambda_2$. Recalling ω_i in (2.2)–(2.3), we define two constants

$$\beta_1 := \inf_{\phi \in H^1(\mathbb{R}^N) \setminus \{0\}} \frac{\int_{\mathbb{R}^N} (|\nabla\phi|^2 + \lambda_2\phi^2)}{\int_{\mathbb{R}^N} \omega_1^2 \phi^2}, \tag{2.6}$$

$$\beta_2 := \inf_{\phi \in H^1(\mathbb{R}^N) \setminus \{0\}} \frac{\int_{\mathbb{R}^N} (|\nabla\phi|^2 + \lambda_1\phi^2)}{\int_{\mathbb{R}^N} \omega_2^2 \phi^2}. \tag{2.7}$$

These two constants were first introduced by Ambrosetti and Colorado [6] in 2006, where the reader can find the significance of these two constants. Furthermore, they proved the following result on the existence of positive (ground state) solutions.

Theorem B ([6, Theorems 1 and 2])

(i) *System (2.1) has a positive radially symmetric solution (U_β, V_β) for any $0 < \beta < \min\{\beta_1, \beta_2\}$.*

(ii) *For any $\beta > \max\{\beta_1, \beta_2\}$, system (2.1) has a positive radially symmetric ground state solution (U_β, V_β) with*

$$E_\beta(U_\beta, V_\beta) = A_\beta < \min\{B_1, B_2\}. \tag{2.8}$$

In the same paper, Ambrosetti and Colorado also suspected that (U_β, V_β) obtained in Theorem B-(i) are also ground state solutions (see [6, Remark 5]). Recently, Ikoma and Tanaka [57] answered this question partially.

Theorem C (see [57, Propositions 2.3 and 2.5, Remark 2.6]) *For any* $0 < \beta < \min\{\beta_1, \beta_2, \sqrt{\mu_1\mu_2}\}$, (U_β, V_β) *obtained in Theorem B-(i) is a ground state solution with*

$$E_\beta(U_\beta, V_\beta) = A_\beta > \max\{B_1, B_2\}. \tag{2.9}$$

Comparing to Theorem B, $\beta < \sqrt{\mu_1\mu_2}$ is assumed in Theorem C because, as pointed out in Chap. 1, that whether $\mathcal{N}_\beta$ is a natural constraint of E_β for $\beta \geq \sqrt{\mu_1\mu_2}$ is *unknown* (see Proposition A).

A basic question is: What are the optimal ranges of the parameter β for the existence of positive (ground state) solutions? This is an open question raised by Sirakov in [80, Remark 4]. Here we are interested in the question concerning *ground state* solutions. In other words, let us define

$$\bar{\beta}_1 := \sup\{\beta' > 0 \mid (2.1) \text{ has a ground state solution for all } 0 < \beta < \beta'\},$$

$$\bar{\beta}_2 := \inf\{\beta' > 0 \mid (2.1) \text{ has a ground state solution for all } \beta > \beta'\}. \tag{2.10}$$

Then both $(0, \bar{\beta}_1)$ and $(\bar{\beta}_2, +\infty)$ are the optimal ranges for the existence of ground state solutions. Our question is: What are the optimal constants $\bar{\beta}_i$, $i = 1, 2$?

Remark that, when $\lambda_1 = \lambda_2$, Theorem A answered this question completely. Since $\beta_1 = \mu_1$ and $\beta_2 = \mu_2$ (the proof will be given later), we see that $\bar{\beta}_i$ are the optimal constants for the existence of ground state solutions, namely $\bar{\beta}_1 = \min\{\beta_1, \beta_2\}$ and $\bar{\beta}_2 = \max\{\beta_1, \beta_2\}$ for $\lambda_1 = \lambda_2$. Indeed, β_i are also the optimal constants for the existence of nontrivial positive solutions for this case $\lambda_1 = \lambda_2$.

A natural question that people are interested in is: For the general case $\lambda_1 \neq \lambda_2$, do we still have $\bar{\beta}_1 = \min\{\beta_1, \beta_2\}$ or $\bar{\beta}_2 = \max\{\beta_1, \beta_2\}$? If so, then β_i are the optimal constants for the existence of ground state solutions for all cases, that is, β_i might have much deeper significance comparing to those pointed out in [6]. Define

$$H_r := \{(u, v) \in H : u, v \text{ are both radially symmetric}\},$$

$$\mathcal{N}_\beta^* := \mathcal{N}_\beta \cap H_r, \quad A_\beta^* := \inf_{(u,v)\in\mathcal{N}_\beta^*} E_\beta(u, v).$$

Without loss of generality, we may assume that $\lambda_1 < \lambda_2$. The following results were proved by Sirakov [80].

Proposition A ([80, Propositions 1.1]) *If A_β (resp. A_β^*) is attained by a couple $(u, v) \in \mathcal{N}_\beta$ (resp. $(u, v) \in \mathcal{N}_\beta^*$), then (u, v) is a critical point of E_β, provided $\beta < \sqrt{\mu_1\mu_2}$.*

Proposition A indicates that, when $\beta < \sqrt{\mu_1\mu_2}$, the existence of ground state solutions is equivalent to that A_β is attained. As pointed out in Chap. 1, whether this conclusion holds or not for the remaining case $\beta \geq \sqrt{\mu_1\mu_2}$ remains open.

Theorem D ([80, Theorem 2.2]) *Suppose $\lambda_1 < \lambda_2$.*

(i) *There exists $\beta' > 0$ such that for any fixed $\beta \in (0, \beta')$, (2.1) has a positive ground state solution (u, v) with $E_\beta(u, v) = A_\beta = A_\beta^*$.*

(ii) *System (2.1) has no nonnegative nontrivial solutions for any $\beta \in [\mu_2, \mu_1]$.*
(iii) *For any $\beta \in [\mu_2, \sqrt{\mu_1\mu_2})$, neither A_β nor A_β^* is attained.*
(iv) *There exists $\beta'' > 0$ such that for any $\beta > \beta''$, (2.1) has a positive ground state solution (u, v) with $E_\beta(u, v) = A_\beta = A_\beta^*$.*

Here we have the following result, which gives the first partial answer to the question of existing optimal constants for the existence of ground state solutions for the general case $\lambda_1 < \lambda_2$ and so improves Theorem D.

Theorem 2.4 *Let $\lambda_1 < \lambda_2$ and $\mu_1 \geq \mu_2$. Then $\beta_2 < \mu_2 \leq \mu_1 < \beta_1$, and*

(i) *there exists small $\delta > 0$ such that (2.1) has no nonnegative nontrivial solutions for any $\beta \in (\mu_2 - \delta, \mu_1 + \delta)$;*
(ii) *for any $\beta \in [\beta_2, \sqrt{\mu_1\mu_2})$, neither A_β nor A_β^* is attained, namely (2.1) has no ground state solutions. Therefore, by Theorem C, it follows that β_2 is an optimal constant for the existence of ground state solutions;*
(iii) *$(U_\beta, V_\beta) \to (0, \omega_2)$ strongly in H as $\beta \uparrow \beta_2$, where (U_β, V_β) is in Theorems B and C;*
(iv) *there exists small $\delta_1 > 0$ such that for any $\beta \in (\beta_2 - \delta_1, \beta_2)$, the ground state solution of (2.1) is unique up to a translation.*

Remark 2.1 Theorem 2.4-(i) and (ii) indicate that (2.1) has *no* ground state solutions for any $\beta \in [\beta_2, \mu_1 + \delta)$.

Remark 2.2 Sirakov [80] gave the precise defintion of β', but *no* information whether β' is an optimal constant or not. Obviously, in the case $\mu_2 \leq \mu_1$, Theorem 2.4 improves Theorem D, and $\beta' \leq \beta_2$ must holds. Besides, we obtain the uniqueness and asymptotic behaviors of ground state solutions as $\beta \uparrow \beta_2$, namely (U_β, V_β) is unique and must converges to the semi-trivial solution $(0, \omega_2)$. It is known that $(0, \omega_2)$ is a semi-trivial solution for all β. Thus we can treat $(0, \omega_2; \beta)$ as a trivial branch of solutions for system (2.1). Our result indicates that, β_2 is actually a bifurcation point, and $(U_\beta, V_\beta; \beta)$ is a nontrivial branch of solutions arising from the trivial branch $(0, \omega_2; \beta)$ at the bifurcation point β_2. This gives a partial answer to an open question raised by Ambrosetti and Corolado [7].

Remark 2.3 If we assume $\lambda_1 > \lambda_2$ and $\mu_1 \leq \mu_2$, we can get a similar theorem. Hence, in the case where $\lambda_1 \neq \lambda_2$ and $(\lambda_2 - \lambda_1)(\mu_2 - \mu_1) \leq 0$, we have $\bar{\beta}_1 = \min\{\beta_1, \beta_2\}$ and so $(0, \min\{\beta_1, \beta_2\})$ is an optimal range for the existence of ground state solutions. This seems to be the first result on this aspect in general case $\lambda_1 \neq \lambda_2$.

Remark 2.4 Theorems 2.1 and 2.4 were published in a joint work with Zou [35], and Theorem 2.2 was published in another joint work with Zou [33]. We remark that Theorem 2.3 is *new* and we did not write it in any articles in the past. On the other hand, recently we proved a non-existence result of nontrivial positive bounded solutions to a more general system (i.e. system (2.1) can be seen as a special case of it) in the half space $\mathbb{R}_+^N := \{x \in \mathbb{R}^N \mid x = (x_1, \ldots, x_N), x_N > 0\}$, where $N \geq 2$ can be arbitrary large and so this system can be of *supercritical* growth; see [28] for details.

We will prove Theorems 2.1 and 2.3 in Sect. 2.2. In Sect. 2.3 we give the proof of Theorem 2.4. Theorem 2.2 will be proved in Chap. 4, where we will study system (1.2) in the critical dimension case $N = 4$. We give some notations here. Throughout this chapter, we denote the norm of $L^p(\mathbb{R}^N)$ by $|u|_p = (\int_{\mathbb{R}^N} |u|^p \, dx)^{\frac{1}{p}}$. Define

$$\|u\|_{\lambda_i}^2 := \int_{\mathbb{R}^N} \left(|\nabla u|^2 + \lambda_i |u|^2 \right) \, dx, \quad i = 1, 2$$

as norms of $H^1(\mathbb{R}^N)$. The norm of H is defined by $\|(u, v)\|^2 := \|u\|_{\lambda_1}^2 + \|v\|_{\lambda_2}^2$.

2.2 Uniqueness of Positive Solutions

First we give the proof of Theorem 2.3 via a simple observation.

Proof (Proof of Theorem 2.3) Let $\lambda_1 = \lambda_2 = \lambda$ and $\mu_1 = \mu_2 = \beta$. Assume that (u, v) is a nontrivial solution of system (2.1) with $u > 0$. By elliptic estimates we see that $u, v \in H^1(\mathbb{R}^N)$. Define $\psi = v/u$. Since $\Delta u + Pu = 0$ and $\Delta v + Pv = 0$, where $P = \beta u^2 + \beta v^2 - \lambda$, we easily conclude that $\nabla \cdot (u^2 \nabla \psi) = 0$. Define cut-off functions $\varphi_R \in C_0^\infty(\mathbb{R}^N)$ such that

$$\varphi_R(x) = \begin{cases} 1, & x \in B_R(0), \\ 0, & x \notin B_{2R}(0), \end{cases} \quad \text{with } |\nabla \varphi| \le \frac{C}{R},$$

where C is independent of R. Then

$$0 = \int_{\mathbb{R}^N} \varphi_R^2 \psi \nabla \cdot (u^2 \nabla \psi) dx = - \int_{\mathbb{R}^N} \nabla(\varphi_R^2 \psi) \cdot (u^2 \nabla \psi) dx$$

$$= - \int_{\mathbb{R}^N} \varphi_R^2 u^2 |\nabla \psi|^2 dx - 2 \int_{\mathbb{R}^N} \varphi_R \psi u^2 \nabla \varphi_R \nabla \psi \, dx,$$

and so

$$\int_{\mathbb{R}^N} \varphi_R^2 u^2 |\nabla \psi|^2 dx \le 2 \int_{\mathbb{R}^N} |\varphi_R u \nabla \psi| |\psi u \nabla \varphi_R| dx$$

$$\le \frac{1}{2} \int_{\mathbb{R}^N} \varphi_R^2 u^2 |\nabla \psi|^2 dx + 2 \int_{\mathbb{R}^N} v^2 |\nabla \varphi_R|^2 dx.$$

Consequently,

$$\int_{|x|\le R} u^2|\nabla\psi|^2\mathrm{d}x \le \int_{\mathbb{R}^N} \varphi_R^2 u^2|\nabla\psi|^2\mathrm{d}x \le 4\int_{\mathbb{R}^N} v^2|\nabla\varphi_R|^2\mathrm{d}x$$

$$\le \frac{4C^2}{R^2}\int_{\mathbb{R}^N} v^2\mathrm{d}x \to 0 \quad \text{as } R \to +\infty.$$

Therefore, $\int_{\mathbb{R}^N} u^2|\nabla\psi|^2\mathrm{d}x = 0$, namely ψ is a non-zero constant. Thus $v = Cu$ for some constant $C \ne 0$. In particular, v does not change sign, and so Theorem 2.3 follows immediately. □

Now let us turn to the proof of Theorem 2.1. Recall that $H_r^1(\mathbb{R}^N)$ is a subspace of $H^1(\mathbb{R}^N)$ that consists of radially symmetric functions. Let ω be the unique radially symmetric positive solution of

$$-\Delta u + u = u^3, \quad u > 0, \quad u \in H^1(\mathbb{R}^N), \tag{2.11}$$

and the corresponding least energy is

$$B := \frac{1}{2}\int_{\mathbb{R}^N}(|\nabla\omega|^2 + \omega^2)\,\mathrm{d}x - \frac{1}{4}\int_{\mathbb{R}^N}\omega^4\,\mathrm{d}x. \tag{2.12}$$

Then it is easy to check that

$$\omega_i(x) = \sqrt{\lambda_i/\mu_i}\,\omega\left(\sqrt{\lambda_i}x\right), \quad B_i = \frac{1}{4}\int_{\mathbb{R}^N}\mu_i\omega_i^4\,\mathrm{d}x = \mu_i^{-1}\lambda_i^{2-N/2}B. \tag{2.13}$$

The following result was proved by Dancer and Wei [47].

Lemma 2.1 ([47, Lemma 2.3]) *When $\beta = \beta_1$, $(u, v) = (\omega_1, 0)$ or $\beta = \beta_2$, $(u, v) = (0, \omega_2)$, the following linearized problem*

$$\begin{cases} \Delta\varphi - \lambda_1\varphi + 3\mu_1 u^2\varphi + \beta v^2\varphi + 2\beta uv\phi = 0, & x \in \mathbb{R}^N, \\ \Delta\phi - \lambda_2\phi + 3\mu_2 v^2\phi + \beta u^2\phi + 2\beta uv\varphi = 0, & x \in \mathbb{R}^N, \\ \varphi, \phi \in H_r^1(\mathbb{R}^N) \end{cases}$$

has exactly a one-dimensional set of solutions.

Now we can give the proof of Theorem 2.1.

Proof (Proof of Theorem 2.1) Let $\lambda_1 = \lambda_2$, and without of loss of generality, we assume $\mu_1 < \mu_2$. Assume by contradiction that there exists $\beta^n \uparrow \mu_1$ as $n \to \infty$, such that (2.1) has a nontrivial nonnegative solution (u_n, v_n) for $\beta = \beta^n$ with

$$\inf_{y \in \mathbb{R}^N} \left\| (u_n(\cdot + y), v_n(\cdot + y)) - (\sqrt{k_{\beta^n}} \omega_0, \sqrt{l_{\beta^n}} \omega_0) \right\| > 0, \quad \forall n \in \mathbb{N}. \tag{2.14}$$

The strong maximum principle gives $u_n, v_n > 0$. By [20], we see that, when $\beta > 0$, any positive solution of (2.1) is radially symmetric decreasing up to a translation. Therefore, we may assume that u_n, v_n are radially symmetric decreasing.

Step 1. We prove that $\|u_n\|_{L^\infty(\mathbb{R}^N)} + \|v_n\|_{L^\infty(\mathbb{R}^N)} \leq C$, where C is a positive constant independent of n.

It is known that

$$-\Delta u \geq \mu_i u^3, \quad u(x) \geq 0, \quad x \in \mathbb{R}^N$$

has no nontrivial solutions if $N \leq 3$. Therefore, this conclusion may follow from a well-known blow up procedure. Since this argument is standard now, we omit the details here, which can be seen in the proof of [47, Lemma 2.4].

Step 2. We show that, for any small $\varepsilon > 0$, there exists $R > 0$ such that

$$u_n(x) + v_n(x) \leq \varepsilon, \quad \forall |x| \geq R, \; \forall n \in \mathbb{N}. \tag{2.15}$$

The details of this proof can also be seen in the proof of [47, Lemma 2.4]. However, since this argument is not trivial, we would like to give the details here for the reader's convenience.

Recalling that u_n, v_n are radially symmetric decreasing, we write $u_n(|x|) = u_n(x)$ and $v_n(|x|) = v_n(x)$ for convenience. Assume that there exists small $\varepsilon > 0$ and $r_n \to +\infty$ such that $u_n(r_n) + v_n(r_n) = \varepsilon$. Define $(\bar{u}_n(r), \bar{v}_n(r)) = (u_n(r + r_n), v_n(r + r_n))$, then

$$\begin{cases} -\bar{u}_n'' - \frac{N-1}{r+r_n} \bar{u}_n' = -\lambda_1 \bar{u}_n + \mu_1 \bar{u}_n^3 + \beta^n \bar{u}_n \bar{v}_n^2, & r > -r_n, \\ -\bar{v}_n'' - \frac{N-1}{r+r_n} \bar{v}_n' = -\lambda_2 \bar{v}_n + \mu_2 \bar{v}_n^3 + \beta^n \bar{v}_n \bar{u}_n^2, & r > -r_n. \end{cases}$$

By elliptic estimates and up to a subsequence, we may assume that $(\bar{u}_n, \bar{v}_n) \to (u, v)$ uniformly in every compact subset of $\mathbb{R}$ as $n \to \infty$, where u, v satisfy

$$\begin{cases} -u'' = -\lambda_1 u + \mu_1 u^3 + \mu_1 u v^2, & r \in \mathbb{R}, \\ -v'' = -\lambda_2 v + \mu_2 v^3 + \mu_1 v u^2, & r \in \mathbb{R}, \end{cases}$$

and $u(0) + v(0) = \varepsilon$, $u, v \geq 0$ are bounded. Since u_n, v_n are both decreasing on $[0, \infty)$, it follows that u, v are both non-increasing on $\mathbb{R}$. Then u, v have limit u_+, v_+ at $+\infty$ and limit u_-, v_- at $-\infty$. Thus, (u_+, v_+) and (u_-, v_-) both satisfy

$$\lambda_1 u = \mu_1 u^3 + \mu_1 u v^2, \quad \lambda_2 v = \mu_2 v^3 + \mu_1 v u^2.$$

Since $\varepsilon > 0$ is small, we have $u_+ = v_+ = 0$ by $u_+ + v_+ \leq \varepsilon$. Since $u_- + v_- \geq \varepsilon$, we may assume that $u_- > 0$, then $\lambda_1 = \mu_1 u_-^2 + \mu_1 v_-^2$. Recall that u and v are non-increasing on $\mathbb{R}$, we see that $u(-\lambda_1 + \mu_1 u^2 + \mu_1 v^2) \leq 0$ on $\mathbb{R}$ and $u(-\lambda_1 + \mu_1 u^2 + \mu_1 v^2) < 0$ on $[0, +\infty)$, which implies that $u'' \geq 0$ on $\mathbb{R}$ and $u'' > 0$ on

$[0, +\infty)$. That is, u is convex on $\mathbb{R}$ and strictly convex on $[0, +\infty)$. This contradicts with $0 \leq u \leq C$, which has been obtained in Step 1. This completes the proof of Step 2.

Step 3. We prove that $\{(u_n, v_n)\}_n$ are uniformly bounded in H.

By (2.15), there exists sufficiently large $R > 0$ such that

$$\max\left\{\mu_1 u_n^2(x) + \beta^n v_n^2(x),\ \mu_2 v_n^2(x) + \beta^n u_n^2(x)\right\} \leq \frac{\lambda_1}{2}, \quad \forall |x| \geq R,\ \forall n \in \mathbb{N}.$$

Since (u_n, v_n) satisfies (2.1), we derive

$$-\Delta u_n(x) + \frac{\lambda_1}{2} u_n(x) \leq 0, \quad -\Delta v_n(x) + \frac{\lambda_1}{2} v_n(x) \leq 0, \quad \forall |x| \geq R,\ \forall n \in \mathbb{N}.$$

Then by a comparison principle, there exists $C > 0$ independent of n such that

$$u_n(x),\ v_n(x) \leq C e^{-\sqrt{\frac{\lambda_1}{2}}|x|}, \quad \forall |x| \geq R,\ \forall n \in \mathbb{N}. \tag{2.16}$$

Define $B(0, R) := \{x \in \mathbb{R}^N : |x| < R\}$. Combining Step 1 with (2.16), it is easily seen that

$$\|u_n\|_{\lambda_1}^2 = \int_{B(0,R)} (\mu_1 u_n^4 + \beta^n u_n^2 v_n^2) + \int_{\mathbb{R}^N \backslash B(0,R)} (\mu_1 u_n^4 + \beta^n u_n^2 v_n^2) \leq C,$$

where C is independent of n. Similarly, $\|v_n\|_{\lambda_2}^2 \leq C$ for C independent of n.

Step 4. We complete the proof via the bifurcation theory.

By Step 3, up to a subsequence, we may assume that $(u_n, v_n) \to (u, v)$ weakly in H and strongly in $L^4(\mathbb{R}^N) \times L^4(\mathbb{R}^N)$ as $\beta^n \uparrow \mu_1$, where $u, v \geq 0$. Then $E'_{\mu_1}(u, v) = 0$ and

$$\lim_{n \to \infty} \|u_n\|_{\lambda_1}^2 = \lim_{n \to \infty} \int_{\mathbb{R}^N} (\mu_1 u_n^4 + \beta^n u_n^2 v_n^2) = \int_{\mathbb{R}^N} (\mu_1 u^4 + \mu_1 u^2 v^2) = \|u\|_{\lambda_1}^2,$$

namely $u_n \to u$ strongly in $H^1(\mathbb{R}^N)$. Similarly, $v_n \to v$ strongly in $H^1(\mathbb{R}^N)$. By Theorem A and (2.13), it easily follows that

$$E_{\mu_1}(u, v) = \lim_{n \to \infty} E_{\beta^n}(u_n, v_n) \geq \lim_{n \to \infty} E_{\beta^n}\left(\sqrt{k_{\beta^n}}\omega_0, \sqrt{l_{\beta^n}}\omega_0\right)$$

$$= E_{\mu_1}(\omega_1, 0) = B_1 > E_{\mu_1}(0, \omega_2) = B_2 > 0. \tag{2.17}$$

Therefore, $(u, v) \neq (0, 0)$. Theorem A-(ii) implies that $u \equiv 0$ or $v \equiv 0$. If $u \equiv 0$, then $v = \omega_2$, a contradiction with (2.17). Hence, $(u, v) = (\omega_1, 0)$. Since $\lambda_1 = \lambda_2$, we have

$$\beta_1 \leq \frac{\int\limits_{\mathbb{R}^N} (|\nabla \omega_1|^2 + \lambda_1 \omega_1^2)}{\int\limits_{\mathbb{R}^N} \omega_1^4} = \mu_1.$$

Moreover, for any $\phi \in H^1(\mathbb{R}^N) \setminus \{0\}$, there holds

$$\frac{\int\limits_{\mathbb{R}^N} (|\nabla \phi|^2 + \lambda_1 \phi^2)}{\int\limits_{\mathbb{R}^N} \omega_1^2 \phi^2} \geq \mu_1 \frac{2\sqrt{B_1} \left(\int\limits_{\mathbb{R}^N} \mu_1 \phi^4\right)^{1/2}}{\left(\int\limits_{\mathbb{R}^N} \mu_1 \omega_1^4\right)^{1/2} \left(\int\limits_{\mathbb{R}^N} \mu_1 \phi^4\right)^{1/2}} = \mu_1.$$

So $\mu_1 = \beta_1$ and $(u_n, v_n, \beta^n) \to (\omega_1, 0, \beta_1)$ is a bifurcation from $(\omega_1, 0, \beta_1)$. By (2.14), $(\sqrt{k_{\beta^n}}\omega_0, \sqrt{l_{\beta^n}}\omega_0, \beta^n) \to (\omega_1, 0, \beta_1)$ is another bifurcation from $(\omega_1, 0, \beta_1)$. By Lemma 2.1, this is a bifurcation from a simple eigenvalue, hence there cannot be two different bifurcations (see [44, 45] or [13, Lemma 3.1]), that is, we get a contradiction. Therefore, there exists small $\delta > 0$ such that for $\mu_1 - \delta < \beta < \mu_1$, $(\sqrt{k_\beta}\omega_0, \sqrt{l_\beta}\omega_0)$ is the unique positive solution to (2.1) up to a translation. $\qquad\square$

2.3 Optimal Parameter Range

In this section, we give the proof of Theorem 2.4. In the sequel we assume that $\lambda_1 < \lambda_2$ and $\mu_1 \geq \mu_2$. Then (2.13) gives $B_1 < B_2$.

Lemma 2.2 *System (2.1) has no nontrivial nonnegative solutions for any $\beta \in [\mu_2, \mu_1]$.*

Proof This result has been pointed out in Theorem D by Sirakov [80], and the proof is very simple. In fact, assume that (2.1) has a nontrivial nonnegative solution (u, v) for some $\beta \in [\mu_2, \mu_1]$, then by the strong maximum principle, we have $u > 0$ and $v > 0$. Multiply the equation for u in (2.1) by v, the equation for v by u, and integrate over $\mathbb{R}^N$, which yields

$$\int\limits_{\mathbb{R}^N} uv[(\lambda_2 - \lambda_1) + (\mu_1 - \beta)u^2 + (\beta - \mu_2)v^2] = 0,$$

a contradiction. $\qquad\square$

Remark 2.5 Theorem D-(iii) is a trivial corollary of Lemma 2.2 and Proposition A. As we will see in the following, the proof of Theorem 2.4 (i)–(ii) seem much more delicate.

Lemma 2.3 $\beta_2 < \mu_2 \le \sqrt{\mu_1\mu_2} \le \mu_1 < \beta_1$.

Proof By (2.7) we have

$$\beta_2 \le \frac{\int\limits_{\mathbb{R}^N} (|\nabla\omega_2|^2 + \lambda_1\omega_2^2)}{\int\limits_{\mathbb{R}^N} \omega_2^4} < \frac{\int\limits_{\mathbb{R}^N} (|\nabla\omega_2|^2 + \lambda_2\omega_2^2)}{\int\limits_{\mathbb{R}^N} \omega_2^4} = \mu_2.$$

On the other hand, it is easy to prove the existence of $\phi_1 \in H^1(\mathbb{R}^N)\backslash\{0\}$ such that

$$\beta_1 = \frac{\int\limits_{\mathbb{R}^N} (|\nabla\phi_1|^2 + \lambda_2\phi_1^2)}{\int\limits_{\mathbb{R}^N} \omega_1^2\phi_1^2}.$$

Then by (2.4) and Hölder inequality, we conclude

$$\beta_1 > \frac{\int\limits_{\mathbb{R}^N} (|\nabla\phi_1|^2 + \lambda_1\phi_1^2)}{\int\limits_{\mathbb{R}^N} \omega_1^2\phi_1^2} \ge \mu_1 \frac{2\sqrt{B_1}\left(\int\limits_{\mathbb{R}^N} \mu_1\phi_1^4\right)^{1/2}}{\left(\int\limits_{\mathbb{R}^N} \mu_1\omega_1^4\right)^{1/2}\left(\int\limits_{\mathbb{R}^N} \mu_1\phi_1^4\right)^{1/2}} = \mu_1.$$

This completes the proof. $\square$

Lemma 2.4 *For any $\beta \in [\beta_2, \sqrt{\mu_1\mu_2})$, there holds $A_\beta \le A_\beta^* \le B_2$. Moreover, $A_{\beta_2} = A_{\beta_2}^* = B_2$.*

Proof Fix any $\beta \in [\beta_2, \sqrt{\mu_1\mu_2})$. As before, it is easy to prove the existence of $\phi_2 \in H_r^1(\mathbb{R}^N)\backslash\{0\}$ such that

$$\beta_2 = J(\phi_2) := \frac{\int\limits_{\mathbb{R}^N} (|\nabla\phi_2|^2 + \lambda_1\phi_2^2)}{\int\limits_{\mathbb{R}^N} \omega_2^2\phi_2^2}, \quad \|\phi_2\|_{\lambda_1} = 1.$$

In fact, ϕ_2 is the first eigenfunction of the following eigenvalue problem

$$-\Delta\phi + \lambda_1\phi = \tau\omega_2^2\phi, \quad \phi \in H^1(\mathbb{R}^N),$$

with the first eigenvalue $\tau_1 = \beta_2$. Hence, for any other $\phi \in H^1(\mathbb{R}^N)$ such that $J(\phi) = \beta_2$, there holds $\phi = C\phi_2$ for some constant C.

If $J(\phi) \leq \beta$ for any $\phi \in H_r^1(\mathbb{R}^N)\backslash\{0\}$, then

$$\frac{\int\limits_{\mathbb{R}^N} (|\nabla\phi|^2 + \lambda_1\phi^2)}{(\int\limits_{\mathbb{R}^N} \phi^4)^{1/2}} \leq \beta \left(\int\limits_{\mathbb{R}^N} \omega_2^4\right)^{1/2} \leq C, \quad \forall \phi \in H_r^1(\mathbb{R}^N)\backslash\{0\}. \qquad (2.18)$$

However, a classical result in [16] proved that the following equation

$$-\Delta u + \lambda_1 u = u^3, \quad u \in H^1(\mathbb{R}^N)$$

has infinitely many sign-changing radially symmetric solutions with energy tending to $+\infty$, which implies that (2.18) cannot hold. Therefore, we may take some $\phi_0 \in H_r^1(\mathbb{R}^N)$ such that $\|\phi_0\|_{\lambda_1} = 1$ and $J(\phi_0) > \beta$. Define

$$u_l := (1 - l)\phi_0 + l\phi_2, \quad 0 \leq l \leq 1.$$

Then $u_l \not\equiv 0$ is radially symmetric for any $0 \leq l \leq 1$, $J(u_0) = J(\phi_0) > \beta$ and $J(u_1) = J(\phi_2) = \beta_2$. Therefore, there exists $0 < l_0 \leq 1$ such that

$$J(u_l) > \beta, \quad \forall 0 \leq l < l_0; \quad J(u_{l_0}) = \beta. \qquad (2.19)$$

Now we let $l \in (0, l_0)$. Note that $(\sqrt{t_l}\, u_l, \sqrt{s_l}\omega_2) \in \mathcal{N}_\beta^*$ for some $t_l, s_l > 0$ is equivalent to $t_l, s_l > 0$ satisfying

$$\int\limits_{\mathbb{R}^N} (|\nabla u_l|^2 + \lambda_1 u_l^2) = t_l \int\limits_{\mathbb{R}^N} \mu_1 u_l^4 + s_l \int\limits_{\mathbb{R}^N} \beta\omega_2^2 u_l^2,$$

$$\int\limits_{\mathbb{R}^N} (|\nabla\omega_2|^2 + \lambda_2\omega_2^2) = s_l \int\limits_{\mathbb{R}^N} \mu_2\omega_2^4 + t_l \int\limits_{\mathbb{R}^N} \beta\omega_2^2 u_l^2 = \int\limits_{\mathbb{R}^N} \mu_2\omega_2^4,$$

that is,

$$t_l = \frac{\int\limits_{\mathbb{R}^N} \mu_2\omega_2^4 \left[\int\limits_{\mathbb{R}^N} (|\nabla u_l|^2 + \lambda_1 u_l^2) - \int\limits_{\mathbb{R}^N} \beta\omega_2^2 u_l^2\right]}{\int\limits_{\mathbb{R}^N} \mu_1 u_l^4 \int\limits_{\mathbb{R}^N} \mu_2\omega_2^4 - \left[\int\limits_{\mathbb{R}^N} \beta\omega_2^2 u_l^2\right]^2},$$

$$s_l = \frac{\int\limits_{\mathbb{R}^N} \mu_1 u_l^4 \int\limits_{\mathbb{R}^N} \mu_2\omega_2^4 - \int\limits_{\mathbb{R}^N} \beta\omega_2^2 u_l^2 \int\limits_{\mathbb{R}^N} (|\nabla u_l|^2 + \lambda_1 u_l^2)}{\int\limits_{\mathbb{R}^N} \mu_1 u_l^4 \int\limits_{\mathbb{R}^N} \mu_2\omega_2^4 - \left[\int\limits_{\mathbb{R}^N} \beta\omega_2^2 u_l^2\right]^2}.$$

Since $\beta < \sqrt{\mu_1 \mu_2}$, we see from Hölder inequality and (2.19) that $t_l > 0$ for all $l \in (0, l_0)$. Since $u_l \to u_{l_0}$ strongly in $H^1(\mathbb{R}^N)$ as $l \uparrow l_0$, it is easy to see from (2.19) that

$$\lim_{l \uparrow l_0}(t_l, s_l) = (0, 1).$$

So $s_l > 0$ for $l_0 - l > 0$ small enough. Recalling that $u_l, \omega_2 \in H_r^1(\mathbb{R}^N)$, we see that $(\sqrt{t_l}\, u_l, \sqrt{s_l}\omega_2) \in \mathcal{N}_\beta^*$ for $l_0 - l > 0$ small enough and then

$$A_\beta \leq A_\beta^* \leq \lim_{l \uparrow l_0} E_\beta(\sqrt{t_l}u_l, \sqrt{s_l}\omega_2) = E_\beta(0, \omega_2) = B_2.$$

Therefore, $A_\beta \leq A_\beta^* \leq B_2$ for any $\beta \in [\beta_2, \sqrt{\mu_1 \mu_2})$.

To finish the proof, it suffices to show that $A_{\beta_2} \geq B_2$. Since $\beta_2 < \sqrt{\mu_1 \mu_2}$ by Lemma 2.3, for any $(u, v) \in \mathcal{N}_{\beta_2}$ and any $0 < \beta < \beta_2$, it is easy to prove the existence of $t_\beta, s_\beta > 0$ such that $(\sqrt{t_\beta}u, \sqrt{s_\beta}v) \in \mathcal{N}_\beta$ and

$$\lim_{\beta \uparrow \beta_2}(t_\beta, s_\beta) = (1, 1).$$

Then

$$\limsup_{\beta \uparrow \beta_2} A_\beta \leq \limsup_{\beta \uparrow \beta_2} E_\beta(\sqrt{t_\beta}u, \sqrt{s_\beta}v) = E_{\beta_2}(u, v), \quad \forall\, (u, v) \in \mathcal{N}_{\beta_2},$$

and so

$$\limsup_{\beta \uparrow \beta_2} A_\beta \leq A_{\beta_2}. \tag{2.20}$$

Consequently, Theorem C gives $A_{\beta_2} \geq B_2$. This completes the proof. $\qquad\square$

Lemma 2.5 *Assume that $\beta_0 \in (0, \sqrt{\mu_1 \mu_2})$ and there exists $(u_0, v_0) \in \mathcal{N}_{\beta_0}^*$ such that $A_{\beta_0}^* = E_{\beta_0}(u_0, v_0)$. Then*

$$A_\beta^* < A_{\beta_0}^* \quad \text{for any } \beta - \beta_0 > 0 \text{ small enough}.$$

Proof Under assumptions in the lemma, it is easy to prove the existence of $t_\beta, s_\beta > 0$ such that $(\sqrt{t_\beta}u_0, \sqrt{s_\beta}v_0) \in \mathcal{N}_\beta^*$ for any $\beta - \beta_0 > 0$ small enough, and $(t_\beta, s_\beta) \to (1, 1)$ as $\beta \to \beta_0$. On the other hand, we note that $(|u_0|, |v_0|) \in \mathcal{N}_{\beta_0}^*$ and $A_{\beta_0}^* = E_{\beta_0}(|u_0|, |v_0|)$. Then by Proposition A one deduces that $(|u_0|, |v_0|)$ is a nontrivial solution of (2.1). Using the strong maximum principle we have $|u_0| > 0$, $|v_0| > 0$, namely $\int_{\mathbb{R}^N} u_0^2 v_0^2 \, dx > 0$. Therefore,

$$A_\beta^* \leq E_\beta(\sqrt{t_\beta}u_0, \sqrt{s_\beta}v_0) = \frac{t_\beta}{4} \int_{\mathbb{R}^N} (|\nabla u_0|^2 + \lambda_1 u_0^2) + \frac{s_\beta}{4} \int_{\mathbb{R}^N} (|\nabla v_0|^2 + \lambda_2 v_0^2)$$

$$= \frac{t_\beta}{4} \int_{\mathbb{R}^N} (\mu_1 u_0^4 + \beta_0 u_0^2 v_0^2) + \frac{s_\beta}{4} \int_{\mathbb{R}^N} (\mu_2 v_0^4 + \beta_0 u_0^2 v_0^2)$$

$$< \frac{t_\beta}{4} \int_{\mathbb{R}^N} (\mu_1 u_0^4 + \beta u_0^2 v_0^2) + \frac{s_\beta}{4} \int_{\mathbb{R}^N} (\mu_2 v_0^4 + \beta u_0^2 v_0^2)$$

$$= \frac{1}{4} \int_{\mathbb{R}^N} (|\nabla u_0|^2 + \lambda_1 u_0^2) + \frac{1}{4} \int_{\mathbb{R}^N} (|\nabla v_0|^2 + \lambda_2 v_0^2)$$

$$= E_{\beta_0}(u_0, v_0) = A_{\beta_0}^*, \quad \text{for any } \beta - \beta_0 > 0 \text{ small enough,}$$

which completes the proof. □

Lemma 2.6 *Let* $\beta_2 < \beta < \sqrt{\mu_1 \mu_2}$. *If* $A_\beta^* < B_2$, *then* (2.1) *has a positive solution* $(u, v) \in \mathcal{N}_\beta^*$ *such that* $E_\beta(u, v) = A_\beta^*$.

Proof Fix any $\beta_2 < \beta < \sqrt{\mu_1 \mu_2}$. Recall that $A_\beta^* \geq A_\beta > 0$. In this proof, we will drop the subscript β for convenience. Note that E is coercive and bounded from below on $\mathcal{N}^*$. Then by the Ekeland variational principle (cf. [81]), there exists a minimizing sequence $\{(u_n, v_n)\} \subset \mathcal{N}^*$ satisfying

$$E(u_n, v_n) \leq A^* + \frac{1}{n}, \tag{2.21}$$

$$E(u', v') \geq E(u_n, v_n) - \frac{1}{n} \|(u_n, v_n) - (u', v')\|, \quad \forall (u', v') \in \mathcal{N}^*. \tag{2.22}$$

Clearly $\{(u_n, v_n)\}$ is bounded in H. Up to a subsequence, we may assume that $(u_n, v_n) \to (u, v)$ weakly in H and strongly in $L^4(\mathbb{R}^N) \times L^4(\mathbb{R}^N)$. Then

$$B_2 > A^* = \frac{1}{4} \lim_{n \to \infty} \int_{\mathbb{R}^N} (\mu_1 u_n^4 + 2\beta u_n^2 v_n^2 + \mu_2 v_n^4)$$

$$= \frac{1}{4} \int_{\mathbb{R}^N} (\mu_1 u^4 + 2\beta u^2 v^2 + \mu_2 v^4) > 0, \tag{2.23}$$

so $(u, v) \neq (0, 0)$.

Step 1. We show that both $u \not\equiv 0$ and $v \not\equiv 0$.

If $u \equiv 0$, then

$$\lim_{n \to \infty} \int_{\mathbb{R}^N} (|\nabla u_n|^2 + \lambda_1 u_n^2) = \lim_{n \to \infty} \int_{\mathbb{R}^N} (\mu_1 u_n^4 + \beta u_n^2 v_n^2) = 0,$$

namely $u_n \to 0$ strongly in $H^1(\mathbb{R}^N)$. Since $\int_{\mathbb{R}^N} (|\nabla v_n|^2 + \lambda_2 v_n^2) = \int_{\mathbb{R}^N} (\mu_2 v_n^4 + \beta u_n^2 v_n^2)$, by Fatou Lemma and (2.4) we have

$$2\sqrt{B_2}\left(\int_{\mathbb{R}^N}\mu_2 v^4\right)^{1/2}\leq\int_{\mathbb{R}^N}(|\nabla v|^2+\lambda_2 v^2)\leq\int_{\mathbb{R}^N}\mu_2 v^4,$$

and so $\int_{\mathbb{R}^N}\mu_2 v^4\geq 4B_2$, a contradiction with (2.23). Hence, $u\not\equiv 0$.

If $v\equiv 0$, similarly we have that $v_n\to 0$ strongly in $H^1(\mathbb{R}^N)$. Define

$$\widetilde{v}_n=\frac{v_n}{|v_n|_4}.$$

Then

$$\|\widetilde{v}_n\|_{\lambda_2}^2=\mu_2|v_n|_4^2+\beta\int_{\mathbb{R}^N}\widetilde{v}_n^2 u_n^2\leq\mu_2|v_n|_4^2+\beta|u_n|_4^2, \tag{2.24}$$

that is, $\widetilde{v}_n$ is uniformly bounded in $H^1(\mathbb{R}^N)$. Passing to a subsequence, $\widetilde{v}_n\rightharpoonup\phi$ weakly in $H^1(\mathbb{R}^N)$. Since $\widetilde{v}_n$ is radially symmetric, we also have $\widetilde{v}_n\to\phi$ strongly in $L^4(\mathbb{R}^N)$, that is, $|\phi|_4=1$ and so $\phi\neq 0$. By (2.24), (2.4), Fatou Lemma and Hölder inequality, one has that

$$2\sqrt{B_2}\left(\int_{\mathbb{R}^N}\mu_2\phi^4\right)^{1/2}\leq\|\phi\|_{\lambda_2}^2\leq\beta\int_{\mathbb{R}^N}u^2\phi^2$$

$$\leq\frac{\beta}{\sqrt{\mu_1\mu_2}}\left(\int_{\mathbb{R}^N}\mu_1 u^4\right)^{1/2}\left(\int_{\mathbb{R}^N}\mu_2\phi^4\right)^{1/2},$$

that is, $\int_{\mathbb{R}^N}\mu_1 u^4\geq\frac{\mu_1\mu_2}{\beta^2}4B_2>4B_2$, a contradiction with (2.23). Hence, $v\not\equiv 0$.

Step 2. We show that $E'_{H_r}(u_n,v_n)\to 0$ as $n\to\infty$.

By Step 1, there exists $C_2>C_1>0$ such that

$$C_1\leq\int_{\mathbb{R}^N}u_n^4\,dx,\int_{\mathbb{R}^N}v_n^4\,dx\leq C_2,\quad\forall n\in\mathbb{N}. \tag{2.25}$$

Thanks to (2.25), the following procedure is a standard argument. For any $(\varphi,\phi)\in H_r$ with $\|\varphi\|,\|\phi\|\leq 1$ and each $n\in\mathbb{N}$, we define h_n and $g_n:\mathbb{R}^3\to\mathbb{R}$ by

$$h_n(t,s,l)=\int_{\mathbb{R}^N}|\nabla(u_n+t\varphi+\frac{s}{2}u_n)|^2+\lambda_1\int_{\mathbb{R}^N}|u_n+t\varphi+\frac{s}{2}u_n|^2$$

$$-\mu_1\int_{\mathbb{R}^N}|u_n+t\varphi+\frac{s}{2}u_n|^4-\beta\int_{\mathbb{R}^N}|u_n+t\varphi+\frac{s}{2}u_n|^2|v_n+t\phi+\frac{l}{2}v_n|^2, \tag{2.26}$$

and

$$g_n(t, s, l) = \int_{\mathbb{R}^N} |\nabla(v_n + t\phi + \frac{l}{2}v_n)|^2 + \lambda_2 \int_{\mathbb{R}^N} |v_n + t\phi + \frac{l}{2}v_n|^2$$

$$- \mu_2 \int_{\mathbb{R}^N} |v_n + t\phi + \frac{l}{2}v_n|^4 - \beta \int_{\mathbb{R}^N} |u_n + t\varphi + \frac{s}{2}u_n|^2 |v_n + t\phi + \frac{l}{2}v_n|^2. \quad (2.27)$$

Let $\mathbf{0} = (0, 0, 0)$. Then $h_n, g_n \in C^1(\mathbb{R}^3, \mathbb{R})$ and $h_n(\mathbf{0}) = g_n(\mathbf{0}) = 0$. Define the matrix

$$F_n := \begin{pmatrix} \frac{\partial h_n}{\partial s}(\mathbf{0}) & \frac{\partial h_n}{\partial l}(\mathbf{0}) \\ \frac{\partial g_n}{\partial s}(\mathbf{0}) & \frac{\partial g_n}{\partial l}(\mathbf{0}) \end{pmatrix}.$$

Then we see from (2.25) that

$$\det(F_n) = \mu_1\mu_2 \int_{\mathbb{R}^N} u_n^4 \, dx \int_{\mathbb{R}^N} v_n^4 \, dx - \beta^2 \left(\int_{\mathbb{R}^N} u_n^2 v_n^2 \, dx \right)^2$$

$$\geq (\mu_1\mu_2 - \beta^2) \int_{\mathbb{R}^N} u_n^4 \, dx \int_{\mathbb{R}^N} v_n^4 \, dx \geq C > 0, \quad (2.28)$$

where C is independent of n. By the implicit function theorem, functions $s_n(t)$ and $l_n(t)$ are well defined and class C^1 on some interval $(-\delta_n, \delta_n)$ for $\delta_n > 0$. Moreover, $s_n(0) = l_n(0) = 0$ and

$$h_n(t, s_n(t), l_n(t)) \equiv 0, \quad g_n(t, s_n(t), l_n(t)) \equiv 0, \quad t \in (-\delta_n, \delta_n).$$

This implies that

$$\begin{cases} s_n'(0) = \frac{1}{\det(F_n)} \left(\frac{\partial g_n}{\partial t}(\mathbf{0}) \frac{\partial h_n}{\partial l}(\mathbf{0}) - \frac{\partial g_n}{\partial l}(\mathbf{0}) \frac{\partial h_n}{\partial t}(\mathbf{0}) \right), \\ l_n'(0) = \frac{1}{\det(F_n)} \left(\frac{\partial g_n}{\partial s}(\mathbf{0}) \frac{\partial h_n}{\partial t}(\mathbf{0}) - \frac{\partial g_n}{\partial t}(\mathbf{0}) \frac{\partial h_n}{\partial s}(\mathbf{0}) \right). \end{cases}$$

On the other hand, since $\{(u_n, v_n)\}$ is bounded in H, we have

$$\left| \frac{\partial h_n}{\partial t}(\mathbf{0}) \right| = 2 \left| \int_{\mathbb{R}^N} (\nabla u_n \nabla \varphi + \lambda_1 u_n \varphi - 2\mu_1 u_n^3 \varphi - \beta u_n v_n^2 \varphi - \beta u_n^2 v_n \phi) \right| \leq C,$$

where C is independent of n. Similarly, $|\frac{\partial g_n}{\partial t}(\mathbf{0})| \leq C$. From (2.25) we also have

$$\left|\frac{\partial h_n}{\partial s}(0)\right|, \quad \left|\frac{\partial h_n}{\partial l}(0)\right|, \quad \left|\frac{\partial g_n}{\partial s}(0)\right|, \quad \left|\frac{\partial g_n}{\partial l}(0)\right| \leq C.$$

Hence, combining these with (2.28), we conclude

$$|s_n'(0)|, \quad |l_n'(0)| \leq C, \tag{2.29}$$

where C is independent of n.

Define $\varphi_{n,t} := u_n + t\varphi + \frac{s_n(t)}{2}u_n$ and $\phi_{n,t} := v_n + t\phi + \frac{l_n(t)}{2}v_n$, then $(\varphi_{n,t}, \phi_{n,t}) \in \mathcal{N}^*$ for $t \in (-\delta_n, \delta_n)$. It follows from (2.22) that

$$E(\varphi_{n,t}, \phi_{n,t}) - E(u_n, v_n) \geq -\frac{1}{n}\left\|\left(t\varphi + \frac{s_n(t)}{2}u_n, t\phi + \frac{l_n(t)}{2}v_n\right)\right\|. \tag{2.30}$$

Note that $E|_{H_r}'(u_n, v_n)(u_n, 0) = E|_{H_r}'(u_n, v_n)(0, v_n) = 0$. By Taylor Expansion we have

$$E(\varphi_{n,t}, \phi_{n,t}) - E(u_n, v_n)$$
$$= E|_{H_r}'(u_n, v_n)\left(t\varphi + \frac{s_n(t)}{2}u_n, t\phi + \frac{l_n(t)}{2}v_n\right) + r(n, t)$$
$$= tE|_{H_r}'(u_n, v_n)(\varphi, \phi) + r(n, t), \tag{2.31}$$

where $r(n, t) = o(\|(t\varphi + \frac{s_n(t)}{2}u_n, t\phi + \frac{l_n(t)}{2}v_n)\|)$ as $t \to 0$. By (2.29) we see that

$$\limsup_{t \to 0}\left\|\left(\varphi + \frac{s_n(t)}{2t}u_n, \phi + \frac{l_n(t)}{2t}v_n\right)\right\| \leq C, \tag{2.32}$$

where C is independent of n. Hence, $r(n, t) = o(t)$. By (2.30), (2.31), (2.32) and letting $t \to 0$, we get that

$$\left|E|_{H_r}'(u_n, v_n)(\varphi, \phi)\right| \leq \frac{C}{n},$$

where C is independent of n. Thus,

$$\lim_{n \to +\infty} E|_{H_r}'(u_n, v_n) = 0. \tag{2.33}$$

Step 3. We show that $(|u|, |v|)$ is a positive solution of system (2.1) such that $E(|u|, |v|) = A^*$.

By Step 2 we have $E|_{H_r}'(u, v) = 0$ and so $(u, v) \in \mathcal{N}^*$. Then we see from (2.23) that $E(u, v) = A^*$. Therefore, $(|u|, |v|) \in \mathcal{N}^*$ and $E(|u|, |v|) = A^*$. Repeating the proof of [80, Proposition 1.1], we have $E|_{H_r}'(|u|, |v|) = 0$. Then by Palais's symmetric criticality principle [75], we see that $E'(|u|, |v|) = 0$. Finally, the maximum principle gives $|u|, |v| > 0$.   □

Now we are in the position to prove Theorem 2.4.

Proof (Proof of Theorem 2.4) (ii) Suppose by contradiction that there exists $\beta_0 \in [\beta_2, \sqrt{\mu_1\mu_2})$ and $(U, V) \in \mathcal{N}_{\beta_0}$ such that $E_{\beta_0}(U, V) = A_{\beta_0}$. By replacing (U, V) by $(|U|, |V|)$ if necessary, we may assume $U, V \geq 0$. Lemma 2.4 gives $A_{\beta_0} \leq B_2$. By Proposition A, we see that (U, V) is a positive ground state solution of (2.1) for $\beta = \beta_0$. By [20] again, we may assume that U and V are both radially symmetric, namely $(U, V) \in \mathcal{N}_{\beta_0}^*$ and so

$$A_{\beta_0}^* \leq E_{\beta_0}(U, V) = A_{\beta_0} \leq A_{\beta_0}^*,$$

which means that $E_{\beta_0}(U, V) = A_{\beta_0}^* \leq B_2$. Define

$$\beta^* = \sup \Big\{ \beta' \in [\beta_0, \sqrt{\mu_1\mu_2}] : \text{for all } \beta \in [\beta_0, \beta'], \ (2.1) \text{ has a positive solution}$$

$$(u_\beta, v_\beta) \in \mathcal{N}_\beta^* \text{ with } E_\beta(u_\beta, v_\beta) = A_\beta^* \Big\}.$$

Lemmas 2.5 and 2.6 indicate that $\beta^* > \beta_0$. Assume by contradiction that $\beta^* < \sqrt{\mu_1\mu_2}$. Then for any $\beta \in (\beta_0, \beta^*)$, (2.1) has a positive solution (u_β, v_β) with $E_\beta(u_\beta, v_\beta) = A_\beta^* < B_2$. Moreover, A_β^* is strictly decreasing with respect to $\beta \in (\beta_0, \beta^*)$. Up to a subsequence, $(u_\beta, v_\beta) \to (u, v)$ weakly in H and strongly in $L^4(\mathbb{R}^N) \times L^4(\mathbb{R}^N)$ as $\beta \uparrow \beta^*$, and $u, v \geq 0$. Then $E'_{\beta^*}(u, v) = 0$ and

$$\lim_{\beta \to \beta^*} \|u_\beta\|_{\lambda_1}^2 = \lim_{\beta \to \beta^*} \int_{\mathbb{R}^N} (\mu_1 u_\beta^4 + \beta u_\beta^2 v_\beta^2) = \int_{\mathbb{R}^N} (\mu_1 u^4 + \beta u^2 v^2) = \|u\|_{\lambda_1}^2,$$

that is, $u_\beta \to u$ strongly in $H^1(\mathbb{R}^N)$. Similarly, $v_\beta \to v$ strongly in $H^1(\mathbb{R}^N)$. Then it is easy to prove that $E_{\beta^*}(u, v) = \lim_{\beta \to \beta^*} A_\beta^* > 0$, and so $(u, v) \neq (0, 0)$. If $u \equiv 0$, then $v = \omega_2$. As before, we define

$$\tilde{u}_\beta = \frac{u_\beta}{|u_\beta|_4}.$$

Similarly as (2.24), we see that $\tilde{u}_\beta$ is uniformly bounded in $H^1(\mathbb{R}^N)$. Passing to a subsequence, $\tilde{u}_\beta \rightharpoonup \phi$ weakly in $H^1(\mathbb{R}^N)$. Since $\tilde{u}_\beta$ is radially symmetric, we also have $\tilde{u}_\beta \to \phi$ strongly in $L^4(\mathbb{R}^N)$, namely $|\phi|_4 = 1$ and $\phi \geq 0$. Since $-\Delta u_\beta + \lambda_1 u_\beta = \mu_1 u_\beta^3 + \beta u_\beta v_\beta^2$, letting $\beta \to \beta^*$ we see that

$$-\Delta\phi + \lambda_1\phi = \beta^* \omega_2^2 \phi, \quad \phi \geq 0 \text{ in } \mathbb{R}^N, \tag{2.34}$$

which implies that $\beta^* = \beta_2$, a contradiction. If $v \equiv 0$, then we may also prove that $\beta^* = \beta_1$, a contradiction with Lemma 2.3. Therefore, $u \not\equiv 0$ and $v \not\equiv 0$, namely $(u, v) \in \mathcal{N}_{\beta^*}^*$ and so $E_{\beta^*}(u, v) \geq A_{\beta^*}^*$. On the other hand, by the same proof of (2.20), we have

$$\limsup_{\beta \uparrow \beta^*} A_\beta^* \leq A_{\beta^*}^* \leq E_{\beta^*}(u, v) = \lim_{\beta \uparrow \beta^*} A_\beta^*.$$

Therefore, $E_{\beta^*}(u, v) = A_{\beta^*}^* = \lim_{\beta \to \beta^*} A_\beta^* < B_2$. By Lemmas 2.5 and 2.6 again, there exists $0 < \varepsilon < \sqrt{\mu_1 \mu_2} - \beta^*$ such that for any $\beta \in [\beta^*, \beta^* + \varepsilon]$, (2.1) has a positive solution (u_β, v_β) with $E_\beta(u_\beta, v_\beta) = A_\beta^*$, which contradicts with the definition of β^*. Therefore, $\beta^* = \sqrt{\mu_1 \mu_2}$. Then by repeating the argument above, we see that (2.1) has a positive solution $(u_{\sqrt{\mu_1 \mu_2}}, v_{\sqrt{\mu_1 \mu_2}})$ for $\beta = \sqrt{\mu_1 \mu_2}$, which contradicts with Lemma 2.2. Therefore, for any $\beta \in [\beta_2, \sqrt{\mu_1 \mu_2})$, A_β is not attained, that is, (2.1) has no ground state solutions for any $\beta \in [\beta_2, \sqrt{\mu_1 \mu_2})$. This proof also implies that A_β^* is not attained for any $\beta \in [\beta_2, \sqrt{\mu_1 \mu_2})$.

(iii) Let (U_β, V_β) be in Theorems B and C. By (2.20) and Theorem C we have

$$\lim_{\beta \uparrow \beta_2} A_\beta = A_{\beta_2} = B_2. \tag{2.35}$$

Assume that there exists a sequence $\beta^n \uparrow \beta_2$ as $n \to \infty$ such that

$$\liminf_{n \to +\infty} \|(U_{\beta^n}, V_{\beta^n}) - (0, \omega_2)\| > 0. \tag{2.36}$$

Up to a subsequence, we may assume that $(U_{\beta^n}, V_{\beta^n}) \to (U, V)$ weakly in H and strongly in $L^4(\mathbb{R}^N) \times L^4(\mathbb{R}^N)$, where $U, V \geq 0$. Similarly, we can prove that $(U_{\beta^n}, V_{\beta^n}) \to (U, V)$ strongly in H, $E'_{\beta_2}(U, V) = 0$ and $E_{\beta_2}(U, V) = A_{\beta_2} = B_2$. Since A_{β_2} is not attained, we have $U \equiv 0$ or $V \equiv 0$. If $V \equiv 0$, then $U \not\equiv 0$. Since $U \geq 0$ is radially symmetric, we see that $U = \omega_1$ and then $E_{\beta_2}(U, V) = E_{\beta_2}(\omega_1, 0) = B_1 < B_2$, a contradiction. Similarly, if $U \equiv 0$, then $V = \omega_2$, which contradicts with (2.36). Therefore, $(U_\beta, V_\beta) \to (0, \omega_2)$ strongly in H as $\beta \uparrow \beta_2$. (i) For $\mu_2 \leq \beta \leq \mu_1$, this result has been proved in Lemma 2.2. Assume that there exists $\beta^n \uparrow \mu_2$ as $n \to \infty$ such that (2.1) has a nonnegative nontrivial solution (u_n, v_n) for $\beta = \beta^n$. By the strong maximum principle, we see that $u_n, v_n > 0$. By [20] again, we may assume that u_n, v_n are radially symmetric decreasing. By a similar argument as the proof of Theorem 2.1, we can prove that $(u_n, v_n) \to (u, v)$ strongly in H and $E'_{\mu_2}(u, v) = 0$. By Remark 2.6, there holds $E_{\beta^n}(u_n, v_n) \geq A_{\beta^n}^* = B_2$ for n sufficiently large, and so $E_{\mu_2}(u, v) \geq B_2 > 0$, that is, $(u, v) \neq (0, 0)$. If $u \equiv 0$, then $v \not\equiv 0$. Since $v \geq 0$ is radially symmetric, we see that $v = \omega_2$. By a similar argument as the proof of (ii), we see that $\mu_2 = \beta_2$, a contradiction with Lemma 2.3. If $v \equiv 0$, then we may prove that $\mu_2 = \beta_1$, also a contradiction. Therefore, $u \not\equiv 0$ and $v \not\equiv 0$, namely (u, v) is a nonnegative nontrivial solution of (2.1) with $\beta = \mu_2$, a contradiction with Lemma 2.2. Therefore, there exists small $\delta_0 > 0$ such that (2.1) has no nonnegative nontrivial solutions for any $\beta \in (\mu_2 - \delta_0, \mu_1]$. Define

$$A'_\beta := \inf_{(u,v)\in\mathcal{N}'_\beta} E_\beta(u,v),$$

where $\mathcal{N}'_\beta := \{(u,v) \in H\backslash\{(0,0)\} : E'_\beta(u,v)(u,v) = 0\}$. Then it is easy to prove that

$$A'_\beta = \inf_{(u,v)\in H\backslash\{(0,0)\}} \max_{t>0} E_\beta(tu, tv).$$

This implies that $A'_\beta > 0$ is non-increasing with respect to β and so

$$\liminf_{\beta\downarrow\mu_1} A'_\beta \geq A'_{\beta_1} > 0.$$

Assume by contradiction that there exists $\beta^n \downarrow \mu_1$ as $n \to \infty$ such that (2.1) has a nonnegative nontrivial solution (u_n, v_n) for $\beta = \beta^n$. Then repeating the proof above, we may prove that $(u_n, v_n) \to (u, v)$ strongly in H and

$$E_{\mu_1}(u, v) = \lim_{\beta^n\downarrow\mu_1} E_{\beta^n}(u_n, v_n) \geq \liminf_{\beta^n\downarrow\mu_1} A'_{\beta^n} > 0.$$

That is, $(u, v) \neq (0, 0)$. Repeating the proof above, we get a contradiction with Lemma 2.2. Therefore, there exists small $\delta \in (0, \delta_0]$ such that (2.1) has no nonnegative nontrivial solutions for any $\beta \in (\mu_2 - \delta, \mu_1 + \delta)$.

(iv) Note that for $\beta < \sqrt{\mu_1\mu_2}$, by a similar argument as Step 3 in the proof of Lemma 2.6, we may assume that all ground state solutions are positive radially symmetric. Assume that there exists $\beta^n \uparrow \beta_2$ as $n \to \infty$ such that (2.1) has a positive radially symmetric ground state solution (u_n, v_n) for $\beta = \beta^n$ with

$$\|(u_n, v_n) - (U_{\beta^n}, V_{\beta^n})\| > 0, \quad \forall n \in \mathbb{N}. \tag{2.37}$$

By a similar argument as the proof of Theorem 2.1, we can prove that $(u_n, v_n) \to (u, v)$ strongly in H and $E'_{\beta_2}(u, v) = 0$. By (2.35), we have

$$E_{\beta_2}(u, v) = \lim_{n\to\infty} E_{\beta^n}(u_n, v_n) = \lim_{n\to\infty} A_{\beta^n} = A_{\beta_2} = B_2,$$

then the proof of (iii) implies that $(u, v) = (0, \omega_2)$, and so (u_n, v_n, β^n) is a bifurcation from $(0, \omega_2, \beta_2)$. Combining (iii) and Lemma 2.1, we get a contradiction just as in the proof of Theorem 2.1. Therefore, there exists small $\delta_1 > 0$ such that for any $\beta \in (\beta_2 - \delta_1, \beta_2)$, (U_β, V_β) is the unique ground state solution of (2.1) up to a translation. This completes the proof. $\qquad\qquad\square$

Remark 2.6 Let $\lambda_1 < \lambda_2$ and $\mu_1 \geq \mu_2$. Lemmas 2.4, 2.6 and Theorem 2.4-(i) imply that

$$A^*_\beta \equiv B_2, \quad \text{for } \beta \in [\beta_2, \sqrt{\mu_1\mu_2}).$$

Then $(\sqrt{t_l}u_l, \sqrt{s_l}\omega_2) \in \mathcal{N}^*_\beta$ constructed in the proof of Lemma 2.4 is indeed a minimizing sequence of A^*_β as $l \to l_0$.

Chapter 3
A BEC System with Dimensions $N = 2, 3$: Sign-Changing Solutions

Abstract As introduced in Chap. 1, we study sign-changing solutions of system (1.2) in a smooth bounded domain Ω, where $\Omega \subset \mathbb{R}^N$ with $N = 2, 3$. In a previous joint work with Lin and Zou, we proved the existence of infinitely many sign-changing solution of (1.2) for the repulsive case $\beta < 0$. In this chapter, we continue our previous research to study the attractive case $\beta > 0$. Precisely, we can obtain multiple sign-changing solutions for $\beta > 0$ small. Furthermore, we can prove the existence of multiple semi-nodal solutions such that one component is sign-changing and the other one is positive. Our proof is purely variational, where a *new* constrained variational problem is introduced. The main tool is the so-called *vector genus* introduced by Tavares and Terracini, which is used to define appropriate minimax values which correspond to sign-changing critical points.

3.1 Main Results

As in Chap. 2, we continue to study the BEC system (1.2). Here for the convenience of notations in the following proof, we rewrite system (1.2) as

$$\begin{cases} -\Delta u_1 + \lambda_1 u_1 = \mu_1 u_1^3 + \beta u_1 u_2^2, & x \in \Omega, \\ -\Delta u_2 + \lambda_2 u_2 = \mu_2 u_2^3 + \beta u_1^2 u_2, & x \in \Omega, \\ u_1|_{\partial\Omega} = u_2|_{\partial\Omega} = 0. \end{cases} \quad (3.1)$$

Differently from Chap. 2, in this chapter we mainly consider the case where Ω is a *smooth bounded domain* in $\mathbb{R}^N (N = 2, 3)$. Our following results hold for any $\lambda_1, \lambda_2 \in (-\lambda_1(\Omega), +\infty)$. As pointed out in Chap. 1, this condition is only used to guarantee that the operators $-\Delta + \lambda_j (j = 1, 2)$ are both positively definite. Thus, without loss of generality, we may assume here that $\lambda_1, \lambda_2 > 0$ just for convenience. In Chap. 2, we mainly study the ground state solutions, which are always *positive* solutions. In this chapter, we focus our attention on studying the existence of *sign-changing* solutions, *semi-nodal* solutions, and *least energy sign-changing* solutions.

© Springer-Verlag Berlin Heidelberg 2015

Z. Chen, *Solutions of Nonlinear Schrödinger Systems*, Springer Theses,
DOI 10.1007/978-3-662-45478-7_3

Definition 3.1 We call a solution (u_1, u_2) is *positive*, if $u_j > 0$ in Ω for $j = 1, 2$; we call a solution (u_1, u_2) *sign-changing* if both u_1 and u_2 change sign in Ω; a solution (u_1, u_2) *semi-nodal* if one component is positive and the other one changes sign. A sign-changing solution (u_1, u_2) is called a *least energy sign-changing solution*, if it attains the least energy among all sign-changing solutions.

As pointed out in Chap. 1, there have been so many interesting results about the existence of (multiple) positive solutions to system (3.1) in the literature. However, to the best of our knowledge, there are few results about sign-changing (and semi-nodal) solutions. The only reference we know is [68], where radially symmetric sign-changing solutions of (1.2) with prescribed zeros were obtained for $\Omega = \mathbb{R}^N$ and $\beta > 0$ sufficiently large. Clearly, their method relies essentially on the radial symmetry of $\mathbb{R}^N$ and so does not work for our case where Ω is a general smooth bounded domain. Note that system (3.1) has infinitely many semi-trivial solutions with the nonzero component sign-changing and might also have multiple positive solutions, and people have to distinguish from all these solutions when seeking nontrivial sign-changing solutions. This fact makes it very challenging to look for sign-changing solutions via variational methods. This is also the main reason why sign-changing solutions of system (3.1) have not been well studied in the literature. Recently, motivated by a recent paper [83], in a joint work with Lin and Zou [26], we could prove the existence of infinitely many sign-changing solutions for the repulsive case $\beta < 0$.

Theorem 3.1 ([26]) *Let* $\beta < 0$. *Then* (3.1) *has infinitely many sign-changing solutions* $(u_{n,1}, u_{n,2})$ *such that*

$$\|u_{n,1}\|_{L^\infty(\Omega)} + \|u_{n,2}\|_{L^\infty(\Omega)} \to +\infty \quad \text{as} \quad n \to +\infty.$$

Remark 3.1 Comparing to those Refs. [13, 48, 55, 72, 87] where infinitely many positive solutions were obtained, here we do not need any symmetry conditions like $\lambda_1 = \lambda_2$ or $\mu_1 = \mu_2$.

In 2005, Lin and Wei [62] proved for $\beta \in (-\infty, \beta_0)$ that (3.1) has a ground state solution which turns out to be a positive solution. Here β_0 is a small positive constant. Since (3.1) has infinitely many sign-changing solutions for any $\beta < 0$, another natural question is whether (3.1) has a least energy sign-changing solution, which has not been studied before. In [26], we also proved the following result.

Theorem 3.2 ([26]) *Let* $\beta < 0$. *Then* (3.1) *has a least energy sign-changing solution* (u_1, u_2). *Moreover, both* u_1 *and* u_2 *have exactly two nodal domains.*

The above two theorems are both concerned with sign-changing solutions. As defined in Definition 3.1, besides positive solutions (see [13, 48, 87]) and sign-changing solutions, it is natural for us to suspect that (3.1) may have semi-nodal solutions. The following result can be also seen in [26].

Theorem 3.3 ([26]) *Let* $\beta < 0$. *Then* (3.1) *has infinitely many semi-nodal solutions* $\{(u_{n,1}, u_{n,2})\}_{n \geq 2}$ *such that*

(1) $u_{n,1}$ *changes sign and* $u_{n,2}$ *is positive;*
(2) $\|u_{n,1}\|_{L^\infty(\Omega)} + \|u_{n,2}\|_{L^\infty(\Omega)} \to +\infty$ *as* $n \to +\infty$;
(3) $u_{n,1}$ *has at most* n *nodal domains. In particular,* $u_{2,1}$ *has exactly two nodal domains, and* $(u_{2,1}, u_{2,2})$ *has the least energy among all nontrivial solutions whose first component changes sign.*

After our paper [26] was submitted, we learn from Professor Z.-Q. Wang that they also obtained infinitely many sign-changing solutions in the repulsive case $\beta < 0$ for a general $m(m \geq 2)$ coupled system in [64]. Remark that our method is quite different from theirs.

In this chapter, we continue our previous work [26] to study the existence of sign-changing solutions and semi-nodal solutions for the remaining case $\beta > 0$. The main idea of the proof is the same as [26]. However, Theorem 2.3 has already told us that (2.1) has *no* semi-nodal solutions when $\mu_1 = \mu_2 = \beta > 0$. This fact indicates that the attractive case $\beta > 0$ is *essentially different* from the repulsive case $\beta < 0$, and here we can only prove the existence of finite multiple sign-changing solutions and semi-nodal solutions for $\beta > 0$ small. Our results are as follows.

Theorem 3.4 *Let* $\Omega \subset \mathbb{R}^N (N = 2, 3)$ *be a smooth bounded domain. Then for any* $k \in \mathbb{N}$, *there exists* $\beta_k > 0$ *such that for any fixed* $\beta \in (0, \beta_k)$, *system* (3.1) *admits at least* k *sign-changing solutions and* k *semi-nodal solutions with the first component sign-changing and the other one positive.*

Theorem 3.5 *Under the same assumption as Theorem 3.4, there exists constant* $\beta_1' \in (0, \beta_1]$ *such that system* (3.1) *has a least energy sign-changing solutions for any fixed* $\beta \in (0, \beta_1')$.

Clearly, we can also prove that (3.1) has at least k semi-nodal solutions with the first component positive and the second one sign-changing for each fixed $\beta \in (0, \beta_k)$. Theorems 3.4 and 3.5 were published in our work [27]. As pointed out in Chap. 1, Sato and Wang [78] proved the existence of $\beta_k > 0$ such that, for each $\beta \in (0, \beta_k)$, (3.1) has at least k nontrivial solutions $(u_{1,i}, u_{2,i})$ with $u_{1,i} > 0$ in Ω ($i = 1, \ldots, k$). Remark that whether $u_{2,i}$ is positive or sign-changing is *not* known in [78]. Clearly our result improves [78]. We should point out that the following proofs are quite different from [78]. In fact, as pointed out before, the main ideas of our proof are inspired by our previous work [26] and a recent paper [83]. In [26], we introduced a *new constrained* problem which is induced from system (3.1). In [83], Tavares and Terracini introduced a new notion of *vector genus* to define appropriate minimax values. More precisely, Tavares and Terracini [83] considered the following general m-coupled system

$$\begin{cases} -\Delta u_j - \mu_j u_j^3 - \beta u_j \sum_{i \neq j} u_i^2 = \lambda_{j,\beta} u_j, \\ u_j \in H_0^1(\Omega), \quad j = 1, \ldots, m, \end{cases} \tag{3.2}$$

where $\beta < 0$, $\mu_j \leq 0$ are all fixed constants (as introduced in Chap. 1, the case $\mu_j < 0$ is called the defocusing case, comparing to the focusing case $\mu_j > 0$ that we study in this thesis). Among other things, they proved that there exist infinitely many $\lambda = (\lambda_{1,\beta}, \ldots, \lambda_{m,\beta}) \in \mathbb{R}^m$ and $u = (u_1, \ldots, u_m) \in H_0^1(\Omega, \mathbb{R}^m)$ such that (u, λ) are sign-changing solutions of (3.2). That is, $\lambda_{j,\beta}$ is not fixed a *priori* but appears as a Lagrange multiplier in (3.2). Therefore, (3.2) is actually a constrained variational problem. This is also an essential difference between the defocusing problem (3.2) and our focusing problem (3.1), where $\lambda_j, \mu_j, \beta > 0$ are all fixed constants. Remark that some arguments in our following proof are borrowed from [26, 83] with modifications. Although some procedures are close to those in [26, 83], for the reader's convenience, we prefer to provide all the necessary details to make the thesis self-contained. Besides, although Theorems 3.4 and 3.5 are both stated in the bounded domain case, our following proofs also work for the entire space case. For this case, we only need to work in the subspace $H_r^1(\mathbb{R}^N)$ (namely every element is radially symmetric) and use the compactness of the Sobolev embedding $H_1^r(\mathbb{R}^N) \hookrightarrow L^4(\mathbb{R}^N)$, then all the following arguments can apply.

We will prove the existence of (least energy) sign-changing solutions in Sect. 3.2. Semi-nodal solutions will be studied in Sect. 3.3. Here we give some notations that will be used in this chapter. Define the norm of $L^p(\Omega)$ as $|u|_p = (\int_\Omega |u|^p \, dx)^{\frac{1}{p}}$, and the norm of $H_0^1(\Omega)$ as $\|u\|^2 = \int_\Omega (|\nabla u|^2 + u^2) \, dx$. Similarly as Chap. 2, we write

$$\|u\|_{\lambda_i}^2 := \int_\Omega (|\nabla u|^2 + \lambda_i u^2) \, dx$$

for convenience. Then $\| \cdot \|_{\lambda_i}$ are norms of $H_0^1(\Omega)$ which are equivalent to $\| \cdot \|$. As before, the norm of $H = H_0^1(\Omega) \times H_0^1(\Omega)$ is $\|(u_1, u_2)\|_H^2 := \|u_1\|_{\lambda_1}^2 + \|u_2\|_{\lambda_2}^2$.

3.2 Sign-Changing Solutions

In this section, we study the existence of sign-changing solutions. Without loss of generality, we assume that $\mu_1 \geq \mu_2$. Let $\beta \in (0, \mu_2)$. Since we are only interested in nontrivial solutions, we define $\widetilde{H} := \{(u_1, u_2) \in H : u_i \neq 0 \text{ for } i = 1, 2\}$, which is an open subset of H. Denote $\mathbf{u} = (u_1, u_2)$ for convenience.

Lemma 3.1 *For any* $\mathbf{u} = (u_1, u_2) \in \widetilde{H}$, *if*

$$\begin{cases} \mu_2 |u_2|_4^4 \|u_1\|_{\lambda_1}^2 - \beta \|u_2\|_{\lambda_2}^2 \int_\Omega u_1^2 u_2^2 \, dx > 0, \\ \mu_1 |u_1|_4^4 \|u_2\|_{\lambda_2}^2 - \beta \|u_1\|_{\lambda_1}^2 \int_\Omega u_1^2 u_2^2 \, dx > 0, \end{cases} \tag{3.3}$$

then system

$$\begin{cases} \|u_1\|_{\lambda_1}^2 = t_1\mu_1|u_1|_4^4 + t_2\beta \int\limits_\Omega u_1^2 u_2^2 \, dx \\ \|u_2\|_{\lambda_2}^2 = t_2\mu_2|u_2|_4^4 + t_1\beta \int\limits_\Omega u_1^2 u_2^2 \, dx \end{cases} \tag{3.4}$$

has a unique solution

$$\begin{cases} t_1(\mathbf{u}) = \dfrac{\mu_2|u_2|_4^4\|u_1\|_{\lambda_1}^2 - \beta\|u_2\|_{\lambda_2}^2 \int\limits_\Omega u_1^2 u_2^2 \, dx}{\mu_1\mu_2|u_1|_4^4|u_2|_4^4 - \beta^2(\int\limits_\Omega u_1^2 u_2^2 \, dx)^2} > 0 \\[4mm] t_2(\mathbf{u}) = \dfrac{\mu_1|u_1|_4^4\|u_2\|_{\lambda_2}^2 - \beta\|u_1\|_{\lambda_1}^2 \int\limits_\Omega u_1^2 u_2^2 \, dx}{\mu_1\mu_2|u_1|_4^4|u_2|_4^4 - \beta^2(\int\limits_\Omega u_1^2 u_2^2 \, dx)^2} > 0. \end{cases} \tag{3.5}$$

Furthermore,

$$\sup_{t_1,t_2\geq 0} E_\beta\left(\sqrt{t_1}u_1, \sqrt{t_2}u_2\right) = E_\beta\left(\sqrt{t_1(\mathbf{u})}u_1, \sqrt{t_2(\mathbf{u})}u_2\right)$$

$$= \frac{1}{4}\left(t_1(\mathbf{u})\|u_1\|_{\lambda_1}^2 + t_2(\mathbf{u})\|u_2\|_{\lambda_2}^2\right)$$

$$= \frac{1}{4}\frac{\mu_2|u_2|_4^4\|u_1\|_{\lambda_1}^4 - 2\beta\|u_1\|_{\lambda_1}^2\|u_2\|_{\lambda_2}^2 \int\limits_\Omega u_1^2 u_2^2 \, dx + \mu_1|u_1|_4^4\|u_2\|_{\lambda_2}^4}{\mu_1\mu_2|u_1|_4^4|u_2|_4^4 - \beta^2(\int\limits_\Omega u_1^2 u_2^2 \, dx)^2}$$

$$\tag{3.6}$$

and $(t_1(\mathbf{u}), t_2(\mathbf{u}))$ is the unique maximum point of $E_\beta(\sqrt{t_1}u_1, \sqrt{t_2}u_2)$.

Proof By (3.3) we have $\mu_1\mu_2|u_1|_4^4|u_2|_4^4 - \beta^2(\int_\Omega u_1^2 u_2^2 dx)^2 > 0$, so $(t_1(\mathbf{u}), t_2(\mathbf{u}))$ defined in (3.5) is the unique solution of (3.4). Note for $t_1, t_2 \geq 0$,

$$f(t_1, t_2) := E_\beta\left(\sqrt{t_1}u_1, \sqrt{t_2}u_2\right) = \frac{1}{2}t_1\|u_1\|_{\lambda_1}^2 + \frac{1}{2}t_2\|u_2\|_{\lambda_2}^2$$

$$- \frac{1}{4}\left(t_1^2\mu_1|u_1|_4^4 + t_2^2\mu_2|u_2|_4^4\right) - \frac{1}{2}t_1 t_2\beta \int\limits_\Omega u_1^2 u_2^2 \, dx$$

$$\leq \left(\frac{t_1}{2}\|u_1\|_{\lambda_1}^2 - \frac{t_1^2}{4}\mu_1|u_1|_4^4\right) + \left(\frac{t_2}{2}\|u_2\|_{\lambda_2}^2 - \frac{t_2^2}{4}\mu_2|u_2|_4^4\right).$$

This yields that $f(t_1, t_2) < 0$ for $\max\{t_1, t_2\} > T$, where T is some positive constant. Consequently, there exists $(\tilde{t}_1, \tilde{t}_2) \in [0, T]^2 \setminus \{(0, 0)\}$ such that

$$f(\tilde{t}_1, \tilde{t}_2) = \sup_{t_1,t_2\geq 0} f(t_1, t_2).$$

To conclude the proof, it suffices to prove that $(\tilde{t}_1, \tilde{t}_2) = (t_1(\mathbf{u}), t_2(\mathbf{u}))$. Note that

$$\sup_{t_1 \geq 0} f(t_1, 0) = \frac{1}{4} \frac{\|u_1\|_{\lambda_1}^4}{\mu_1 |u_1|_4^4}.$$

Recalling the expression of $f(t_1(\mathbf{u}), t_2(\mathbf{u}))$ in (3.6), by a direct computation we derive from (3.3) that

$$f(t_1(\mathbf{u}), t_2(\mathbf{u})) - \sup_{t_1 \geq 0} f(t_1, 0) = \frac{(\mu_1 |u_1|_4^4 \|u_2\|_{\lambda_2}^2 - \beta \|u_1\|_{\lambda_1}^2 \int_\Omega u_1^2 u_2^2 \, dx)^2}{4\mu_1 |u_1|_4^4 [\mu_1 \mu_2 |u_1|_4^4 |u_2|_4^4 - \beta^2 (\int_\Omega u_1^2 u_2^2 \, dx)^2]} > 0.$$

Similarly we also have $f(t_1(\mathbf{u}), t_2(\mathbf{u})) - \sup_{t_2 \geq 0} f(0, t_2) > 0$, so $\tilde{t}_1 > 0$ and $\tilde{t}_2 > 0$. Then by $\frac{\partial}{\partial t_1} f(t_1, t_2)|_{(\tilde{t}_1, \tilde{t}_2)} = \frac{\partial}{\partial t_2} f(t_1, t_2)|_{(\tilde{t}_1, \tilde{t}_2)} = 0$ we conclude that $(\tilde{t}_1, \tilde{t}_2)$ satisfies (3.4), so $(\tilde{t}_1, \tilde{t}_2) = (t_1(\mathbf{u}), t_2(\mathbf{u}))$. $\square$

Now we define

$$\mathscr{M}^* := \left\{ \mathbf{u} \in H \mid 1/2 < |u_1|_4^4 < 2, \ 1/2 < |u_2|_4^4 < 2 \right\}; \tag{3.7}$$

$$\mathscr{M}_\beta^* := \left\{ \mathbf{u} \in \mathscr{M}^* \mid \mathbf{u} \text{ satisfies } (3.3) \right\};$$

$$\mathscr{M} := \{ \mathbf{u} \in H \mid |u_1|_4 = 1, \ |u_2|_4 = 1 \}, \quad \mathscr{M}_\beta := \mathscr{M} \cap \mathscr{M}_\beta^*; \tag{3.8}$$

$$\mathscr{M}_\beta^{**} := \left\{ \mathbf{u} \in \mathscr{M}^* \; \middle| \; \begin{array}{l} \mu_2 \|u_1\|_{\lambda_1}^2 - \beta \|u_2\|_{\lambda_2}^2 \int_\Omega u_1^2 u_2^2 \, dx > 0 \\ \mu_1 \|u_2\|_{\lambda_2}^2 - \beta \|u_1\|_{\lambda_1}^2 \int_\Omega u_1^2 u_2^2 \, dx > 0 \end{array} \right\}.$$

Then it is easy to see that $\mathscr{M}_\beta \neq \emptyset$ and $\mathscr{M}_\beta = \mathscr{M} \cap \mathscr{M}_\beta^{**}$. Clearly $\mathscr{M}^*$, $\mathscr{M}_\beta^*$, and $\mathscr{M}_\beta^{**}$ are all open subsets of H, while $\mathscr{M}$ is closed. Noting that $\mu_1 \mu_2 - \beta^2 (\int_\Omega u_1^2 u_2^2 \, dx)^2 > 0$ for any $\mathbf{u} \in \mathscr{M}_\beta^{**}$, as in [26], we define a new functional $J_\beta : \mathscr{M}_\beta^{**} \to (0, +\infty)$ by (compare with (3.6))

$$J_\beta(\mathbf{u}) := \frac{1}{4} \frac{\mu_2 \|u_1\|_{\lambda_1}^4 - 2\beta \|u_1\|_{\lambda_1}^2 \|u_2\|_{\lambda_2}^2 \int_\Omega u_1^2 u_2^2 \, dx + \mu_1 \|u_2\|_{\lambda_2}^4}{\mu_1 \mu_2 - \beta^2 (\int_\Omega u_1^2 u_2^2 \, dx)^2}.$$

A direct computation yields $J_\beta \in C^1(\mathscr{M}_\beta^{**}, (0, +\infty))$. Moreover, since any $\mathbf{u} \in \mathscr{M}_\beta$ is an interior point of $\mathscr{M}_\beta^{**}$, by (3.5) we can prove via a direct calculation that

$$J_\beta'(\mathbf{u})(\varphi, 0) = t_1(\mathbf{u}) \int_\Omega (\nabla u_1 \nabla \varphi + \lambda_1 u_1 \varphi) \, dx - t_1(\mathbf{u}) t_2(\mathbf{u}) \beta \int_\Omega u_1 u_2^2 \varphi \, dx, \tag{3.9}$$

$$J_\beta'(\mathbf{u})(0, \psi) = t_2(\mathbf{u}) \int_\Omega (\nabla u_2 \nabla \psi + \lambda_2 u_2 \psi) \, dx - t_1(\mathbf{u}) t_2(\mathbf{u}) \beta \int_\Omega u_1^2 u_2 \psi \, dx$$

(3.10)

hold for any $\mathbf{u} \in \mathscr{M}_\beta$ and $\varphi, \psi \in H_0^1(\Omega)$ (remark that (3.9)–(3.10) do not hold for $\mathbf{u} \in \mathscr{M}_\beta^{**} \setminus \mathscr{M}_\beta$). Besides, Lemma 3.1 yields

$$J_\beta(u_1, u_2) = \sup_{t_1, t_2 \geq 0} E_\beta\left(\sqrt{t_1} u_1, \sqrt{t_2} u_2\right), \quad \forall (u_1, u_2) \in \mathscr{M}_\beta.$$

(3.11)

To obtain nontrivial solutions of (3.1), we turn to study the functional J_β restricted to $\mathscr{M}_\beta$, which is a constrained variational problem with two constraints.

Remark 3.2 To obtain nontrivial solutions of (3.1), in many papers (see [48, 61, 62, 80, 87] for example), people usually turn to study nontrivial critical points of E_β under the following Nehari manifold type constraint

$$\left\{(u_1, u_2) \in \tilde{H} : E_\beta'(u_1, u_2)(u_1, 0) = E_\beta'(u_1, u_2)(0, u_2) = 0\right\},$$

which is actually a *natural constraint* for any $\beta < \sqrt{\mu_1 \mu_2}$ (see Proposition A in Chap. 2). To the best of our knowledge, our natural idea (i.e., to obtain nontrivial solutions of (3.1) by studying $J_\beta|_{\mathscr{M}_\beta}$) introduced in [26] is *new*, and has *never* been used for (3.1) in the literature.

Different from the repulsive case $\beta < 0$ studied in [26], here we need to define new sets

$$\mathscr{N}_b^* := \left\{\mathbf{u} \in \mathscr{M}^* : \|u_1\|_{\lambda_1}^2, \|u_2\|_{\lambda_2}^2 < b\right\}, \quad \mathscr{N}_b := \mathscr{N}_b^* \cap \mathscr{M}.$$

(3.12)

Fix any $k \in \mathbb{N}$. Our goal is to prove the existence of $\beta_k > 0$ such that (3.1) has at least k sign-changing solutions for any $\beta \in (0, \beta_k)$. To do this, we let W_{k+1} be a $k+1$ dimensional subspace of $H_0^1(\Omega)$ which contains an element φ_0 satisfying $\varphi_0 > 0$ in Ω. Then we can find $\bar{b} > 0$ such that

$$\|u\|_{\lambda_1}^2, \|u\|_{\lambda_2}^2 < \bar{b}, \quad \forall u \in W_{k+1} \text{ satisfying } |u|_4^4 < 2.$$

(3.13)

Fix a $b > 0$ such that

$$b^2 > (2 + \mu_1/\mu_2)\bar{b}^2.$$

(3.14)

Then $\mathscr{N}_{\bar{b}}^* \subset \mathscr{N}_b^*$ and $\mathscr{N}_{\bar{b}} \subset \mathscr{N}_b$. Recalling the Sobolev inequality

$$\|u\|_{\lambda_i}^2 \geq \mathscr{C} |u|_4^2, \quad \forall u \in H_0^1(\Omega), \quad i = 1, 2,$$

(3.15)

where $\mathscr{C}$ is a positive constant, we have the following lemma.

Lemma 3.2 *There exist $\beta_0 \in (0, \mu_2)$ and constants $C_1 > C_0 > 0$ such that for any $\beta \in (0, \beta_0)$ there hold $\mathcal{N}_b^* \subset \mathcal{M}_\beta^* \cap \mathcal{M}_\beta^{**}$ and*

$$C_0 \leq t_1(\mathbf{u}), \quad t_2(\mathbf{u}) \leq C_1, \quad \forall \mathbf{u} \in \mathcal{N}_b^*.$$

Proof Define $\beta_0 := \frac{\mu_2 \mathscr{C}}{8b}$ and let $\beta \in (0, \beta_0)$. For any $\mathbf{u} = (u_1, u_2) \in \mathcal{N}_b^*$, we derive from (3.7) and (3.15) that $\int_\Omega u_1^2 u_2^2 \, dx \leq |u_1|_4^2 |u_2|_4^2 < 2$ and $\|u_i\|_{\lambda_i}^2 \geq \mathscr{C}/\sqrt{2}$. Hence, it is easy to check that

$$\mu_2 |u_2|_4^4 \|u_1\|_{\lambda_1}^2 - \beta \|u_2\|_{\lambda_2}^2 \int_\Omega u_1^2 u_2^2 \, dx \geq \frac{\mu_2 \mathscr{C}}{2\sqrt{2}} - 2b\beta_0 \geq \frac{\mu_2 \mathscr{C}}{16};$$

$$\mu_1 |u_1|_4^4 \|u_2\|_{\lambda_2}^2 - \beta \|u_1\|_{\lambda_1}^2 \int_\Omega u_1^2 u_2^2 \, dx \geq \frac{\mu_2 \mathscr{C}}{16};$$

$$\mu_2 \|u_1\|_{\lambda_1}^2 - \beta \|u_2\|_{\lambda_2}^2 \int_\Omega u_1^2 u_2^2 \, dx \geq \frac{\mu_2 \mathscr{C}}{16};$$

$$\mu_1 \|u_2\|_{\lambda_2}^2 - \beta \|u_1\|_{\lambda_1}^2 \int_\Omega u_1^2 u_2^2 \, dx \geq \frac{\mu_2 \mathscr{C}}{16};$$

$$\mu_1 \mu_2 - \beta^2 \left(\int_\Omega u_1^2 u_2^2 \, dx \right)^2 \geq \frac{\mu_2^2 \mathscr{C}^2}{2^8} \cdot \frac{1}{\|u_1\|_{\lambda_1}^2 \|u_2\|_{\lambda_2}^2} \geq \frac{\mu_2^2 \mathscr{C}^2}{2^8 b^2};$$

$$\mu_1 \mu_2 |u_1|_4^4 |u_2|_4^4 - \beta^2 \left(\int_\Omega u_1^2 u_2^2 \, dx \right)^2 \geq \frac{\mu_2^2 \mathscr{C}^2}{2^8 b^2}.$$

Consequently, $\mathbf{u} \in \mathcal{M}_\beta^* \cap \mathcal{M}_\beta^{**}$. Moreover, combining these with (3.5) we have

$$t_i(\mathbf{u}) \geq \frac{\mu_2 \mathscr{C}}{2^4} \cdot \frac{1}{\mu_1 \mu_2 |u_1|_4^4 |u_2|_4^4} \geq \frac{\mathscr{C}}{2^6 \mu_1}, \quad t_i(\mathbf{u}) \leq \frac{2^9 b^3}{\mu_2^2 \mathscr{C}^2} \mu_1, \quad i = 1, 2.$$

This completes the proof. $\square$

Lemma 3.3 *There exist $\beta_k \in (0, \beta_0]$ and $d_k > 0$ such that*

$$\inf_{\mathbf{u} \in \partial \mathcal{N}_b} J_\beta(\mathbf{u}) \geq d_k > \sup_{\mathbf{u} \in \mathcal{N}_{\tilde{b}}} J_\beta(\mathbf{u}), \quad \forall \beta \in (0, \beta_k). \tag{3.16}$$

Proof This proof is inspired by [78]. Define two functionals

$$I_i(u_i) := \frac{1}{4\mu_i} \|u_i\|_{\lambda_i}^4, \quad i = 1, 2.$$

Then for any $\mathbf{u} \in \overline{\mathcal{N}_b}$ and $\beta \in (0, \beta_0)$, we have

$$|J_\beta(\mathbf{u}) - I_1(u_1) - I_2(u_2)|$$

$$= \frac{\beta \left| \beta (\int_\Omega u_1^2 u_2^2 \, dx)^2 \sum_{i=1}^2 \|u_i\|_{\lambda_i}^4 / \mu_i - 2\|u_1\|_{\lambda_1}^2 \|u_2\|_{\lambda_2}^2 \int_\Omega u_1^2 u_2^2 \, dx \right|}{4[\mu_1 \mu_2 - \beta^2 (\int_\Omega u_1^2 u_2^2 \, dx)^2]} \le C\beta,$$

where $C > 0$ is independent of $\mathbf{u} \in \overline{\mathcal{N}_b}$ and $\beta \in (0, \beta_0)$. Consequently

$$\sup_{\mathbf{u} \in \mathcal{N}_b^*} J_\beta(\mathbf{u}) \le \sup_{\mathbf{u} \in \mathcal{N}_b^*} (I_1(u_1) + I_2(u_2)) + C\beta \le \frac{\bar{b}^2}{4\mu_1} + \frac{\bar{b}^2}{4\mu_2} + C\beta,$$

$$\inf_{\mathbf{u} \in \partial \mathcal{N}_b} J_\beta(\mathbf{u}) \ge \inf_{\mathbf{u} \in \partial \mathcal{N}_b} (I_1(u_1) + I_2(u_2)) - C\beta \ge \frac{b^2}{4\mu_1} - C\beta.$$

Recalling (3.14), we let $\beta_k = \min\{\frac{b^2}{8\mu_1 C}, \beta_0\}$ and $d_k = \frac{b^2}{4\mu_1} - C\beta_k$, then (3.16) follows immediately. $\qquad\qquad\square$

In the following, we always let $(i, j) = (1, 2)$ or $(i, j) = (2, 1)$. Recalling (3.15) and Lemma 3.2, we can take β_k smaller if necessary such that, for any $\beta \in (0, \beta_k)$ and $\mathbf{u} \in \mathcal{N}_b^*$, there holds

$$\|v\|_{\lambda_i}^2 - \beta t_j(\mathbf{u}) \int_\Omega u_j^2 v^2 \, dx \ge \frac{1}{2}\|v\|_{\lambda_i}^2, \quad \forall v \in H_0^1(\Omega), \quad i = 1, 2. \tag{3.17}$$

Clearly (3.17) indicates that the operators $-\Delta + \lambda_i - \beta t_j(\mathbf{u})u_j^2$ are positively definite in $H_0^1(\Omega)$.

From now on, we fix any $\beta \in (0, \beta_k)$, and we will prove that (3.1) has at least k sign-changing solutions. For any $\mathbf{u} = (u_1, u_2) \in \mathcal{N}_b^*$, we let $\tilde{w}_i \in H_0^1(\Omega)$ be the unique solution of the following linear equation

$$-\Delta \tilde{w}_i + \lambda_i \tilde{w}_i - \beta t_j(\mathbf{u})u_j^2 \tilde{w}_i = \mu_i t_i(\mathbf{u})u_i^3, \quad \tilde{w}_i \in H_0^1(\Omega). \tag{3.18}$$

Since $|u_i|_4^4 > 1/2$, so $\tilde{w}_i \neq 0$ and we derive from (3.17) that

$$\int_\Omega u_i^3 \tilde{w}_i \, dx = \frac{1}{\mu_i t_i(\mathbf{u})} \left(\|\tilde{w}_i\|_{\lambda_i}^2 - \beta t_j(\mathbf{u}) \int_\Omega u_j^2 \tilde{w}_i^2 \, dx \right) \ge \frac{1}{2\mu_i t_i(\mathbf{u})} \|\tilde{w}_i\|_{\lambda_i}^2 > 0.$$

Then we can define

$$w_i = \alpha_i \tilde{w}_i, \quad \text{where} \quad \alpha_i = \frac{1}{\int\limits_\Omega u_i^3 \tilde{w}_i \, dx} > 0. \tag{3.19}$$

Consequently, w_i is the unique solution of the following problem

$$\begin{cases} -\Delta w_i + \lambda_i w_i - \beta t_j(\mathbf{u}) u_j^2 w_i = \alpha_i \mu_i t_i(\mathbf{u}) u_i^3, \quad w_i \in H_0^1(\Omega), \\ \int\limits_\Omega u_i^3 w_i \, dx = 1. \end{cases} \tag{3.20}$$

Now we define an operator $K = (K_1, K_2) : \mathcal{N}_b^* \to H$ by

$$K(\mathbf{u}) = (K_1(\mathbf{u}), K_2(\mathbf{u})) := \mathbf{w} = (w_1, w_2). \tag{3.21}$$

Define the transformations

$$\sigma_i : H \to H \quad \text{by} \quad \sigma_1(u_1, u_2) := (-u_1, u_2), \quad \sigma_2(u_1, u_2) := (u_1, -u_2). \tag{3.22}$$

Then it is easy to check that

$$K(\sigma_i(\mathbf{u})) = \sigma_i(K(\mathbf{u})), \quad i = 1, 2. \tag{3.23}$$

Lemma 3.4 $K \in C^1(\mathcal{N}_b^*, H)$.

Proof Similarly as in [83], it suffices to apply the implicit function theorem to the C^1 map

$$\Psi : \mathcal{N}_b^* \times H_0^1(\Omega) \times \mathbb{R} \to H_0^1(\Omega) \times \mathbb{R}, \quad \text{where}$$

$$\Psi(\mathbf{u}, v, \alpha) = \left(v - (-\Delta + \lambda_i)^{-1} \left(\beta t_j(\mathbf{u}) u_j^2 v + \alpha \mu_i t_i(\mathbf{u}) u_i^3 \right), \int\limits_\Omega u_i^3 v \, dx - 1 \right).$$

Remark that (3.20) holds if and only if $\Psi(\mathbf{u}, w_i, \alpha_i) = (0, 0)$. By computing the derivative of Ψ with respect to (v, α) at the point $(\mathbf{u}, w_i, \alpha_i)$ in the direction $(\bar{w}, \bar{\alpha})$, we obtain a map $\Phi : H_0^1(\Omega) \times \mathbb{R} \to H_0^1(\Omega) \times \mathbb{R}$ given by

$$\Phi(\bar{w}, \bar{\alpha}) := D_{v,\alpha} \Psi(\mathbf{u}, w_i, \alpha_i)(\bar{w}, \bar{\alpha})$$

$$= \left(\bar{w} - (-\Delta + \lambda_i)^{-1} \left(\beta t_j(\mathbf{u}) u_j^2 \bar{w} + \bar{\alpha} \mu_i t_i(\mathbf{u}) u_i^3 \right), \int\limits_\Omega u_i^3 \bar{w} \, dx \right).$$

If $\Phi(\bar{w}, \bar{\alpha}) = (0, 0)$, then we multiply the equation

$$-\Delta \bar{w} + \lambda_i \bar{w} - \beta t_j(\mathbf{u}) u_j^2 \bar{w} = \bar{\alpha} \mu_i t_i(\mathbf{u}) u_i^3$$

by $\bar{w}$ and use (3.17) to obtain

$$\frac{1}{2} \|\bar{w}\|_{\lambda_i}^2 \leq \bar{\alpha} \mu_i t_i(\mathbf{u}) \int_{\Omega} u_i^3 \bar{w} \, dx = 0.$$

So $\bar{w} = 0$ and then $\bar{\alpha} \mu_i t_i(\mathbf{u}) u_i^3 \equiv 0$ in Ω. Since $\mu_i > 0$, $t_i(\mathbf{u}) > 0$ and $|u_i|_4 \geq 1/2$, we conclude $\bar{\alpha} = 0$. Hence Φ is injective.

On the other hand, for any $(f, c) \in H_0^1(\Omega) \times \mathbb{R}$, let $v_1, v_2 \in H_0^1(\Omega)$ be solutions of the linear problems

$$- \Delta v_1 + \lambda_i v_1 - \beta t_j(\mathbf{u}) u_j^2 v_1 = (-\Delta + \lambda_i) f,$$
$$- \Delta v_2 + \lambda_i v_2 - \beta t_j(\mathbf{u}) u_j^2 v_2 = \mu_i t_i(\mathbf{u}) u_i^3.$$

Since $|u_i|_4 > 1/2$, so $v_2 \neq 0$ and then (3.17) yields $\int_{\Omega} u_i^3 v_2 \, dx > 0$. Let $\alpha_0 = (c - \int_{\Omega} u_i^3 v_1 \, dx) / \int_{\Omega} u_i^3 v_2 \, dx$, then $\Phi(v_1 + \alpha_0 v_2, \alpha_0) = (f, c)$. Hence Φ is surjective, that is, Φ is a bijective map. Thus, the proof is complete. $\qquad\square$

Lemma 3.5 *Assume that* $\{\mathbf{u}_n = (u_{n,1}, u_{n,2}) : n \geq 1\} \subset \mathcal{N}_b$. *Then there exists* $\mathbf{w} \in H$ *such that, up to a subsequence,* $\mathbf{w}_n := K(\mathbf{u}_n) \to \mathbf{w}$ *strongly in* H.

Proof Up to a subsequence, we may assume that $\mathbf{u}_n \rightharpoonup \mathbf{u} = (u_1, u_2)$ weakly in H and so $u_{n,i} \to u_i$ strongly in $L^4(\Omega)$, which implies $|u_i|_4 = 1$. Moreover, by Lemma 3.2 we may assume $t_i(\mathbf{u}_n) \to t_i > 0$. Recall that $w_{n,i} = \alpha_{n,i} \tilde{w}_{n,i}$, where $\alpha_{n,i}$ and $\tilde{w}_{n,i}$ are seen in (3.18)–(3.19). By (3.17)–(3.18) we have

$$\frac{1}{2} \|\tilde{w}_{n,i}\|_{\lambda_i}^2 \leq \mu_i t_i(\mathbf{u}_n) \int_{\Omega} u_{n,i}^3 \tilde{w}_{n,i} \, dx \leq C |\tilde{w}_{n,i}|_4 \leq C \|\tilde{w}_{n,i}\|_{\lambda_i},$$

which implies that $\{\tilde{w}_{n,i} : n \geq 1\}$ are bounded in $H_0^1(\Omega)$. Up to a subsequence, we may assume that $\tilde{w}_{n,i} \to \tilde{w}_i$ weakly in $H_0^1(\Omega)$ and strongly in $L^4(\Omega)$. Then by (3.18) it is standard to prove that $\tilde{w}_{n,i} \to \tilde{w}_i$ strongly in $H_0^1(\Omega)$. Moreover, $\tilde{w}_i$ satisfies

$$-\Delta \tilde{w}_i + \lambda_i \tilde{w}_i - \beta t_j u_j^2 \tilde{w}_i = \mu_i t_i u_i^3.$$

Since $|u_i|_4 = 1$, we have $\tilde{w}_i \neq 0$ and then $\int_{\Omega} u_i^3 \tilde{w}_i \, dx > 0$, which implies that

$$\lim_{n \to \infty} \alpha_{n,i} = \lim_{n \to \infty} \frac{1}{\int_{\Omega} u_{n,i}^3 \tilde{w}_{n,i} \, dx} = \frac{1}{\int_{\Omega} u_i^3 \tilde{w}_i \, dx} =: \alpha_i.$$

Therefore, $w_{n,i} = \alpha_{n,i} \tilde{w}_{n,i} \to \alpha_i \tilde{w}_i =: w_i$ strongly in $H_0^1(\Omega)$. $\qquad\square$

To continue our proof, we need to use *vector genus* introduced by [83] to define proper minimax energy levels. Recalling (3.8) and (3.22), as in [83] we consider the class of sets

$$\mathscr{F} = \{A \subset \mathscr{M} \mid A \text{ is closed and } \sigma_i(\mathbf{u}) \in A, \quad \forall \mathbf{u} \in A, \quad i = 1, 2\},$$

and, for each $A \in \mathscr{F}$ and $k_1, k_2 \in \mathbb{N}$, the class of functions

$$F_{(k_1, k_2)}(A) = \left\{ f = (f_1, f_2) : A \to \prod_{i=1}^{2} \mathbb{R}^{k_i - 1} \,\middle|\, \begin{array}{l} f_i : A \to \mathbb{R}^{k_i - 1} \text{ continuous}, \\ f_i(\sigma_i(\mathbf{u})) = -f_i(\mathbf{u}) \text{ for each } i, \\ f_i(\sigma_j(\mathbf{u})) = f_i(\mathbf{u}) \text{ for } j \neq i \end{array} \right\}.$$

Here we denote $\mathbb{R}^0 := \{0\}$. Let us recall vector genus from [83].

Definition 3.2 (Vector genus, see [83]) Let $A \in \mathscr{F}$ and take any $k_1, k_2 \in \mathbb{N}$. We say that $\gamma(A) \geq (k_1, k_2)$ if for every $f \in F_{(k_1, k_2)}(A)$ there exists $\mathbf{u} \in A$ such that $f(\mathbf{u}) = (f_1(\mathbf{u}), f_2(\mathbf{u})) = (0, 0)$. We denote

$$\Gamma^{(k_1, k_2)} := \{A \in \mathscr{F} \mid \gamma(A) \geq (k_1, k_2)\}.$$

Lemma 3.6 (see [83]) *With the previous notations, the following properties hold.*

(i) *Take $A_1 \times A_2 \subset \mathscr{M}$ and let $\eta_i : S^{k_i - 1} := \{x \in \mathbb{R}^{k_i} : |x| = 1\} \to A_i$ be a homeomorphism such that $\eta_i(-x) = -\eta_i(x)$ for every $x \in S^{k_i - 1}$, $i = 1, 2$. Then $\overline{A_1 \times A_2} \in \Gamma^{(k_1, k_2)}$.*

(ii) *We have $\overline{\eta(A)} \in \Gamma^{(k_1, k_2)}$ whenever $A \in \Gamma^{(k_1, k_2)}$ and a continuous map $\eta : A \to \mathscr{M}$ is such that $\eta \circ \sigma_i = \sigma_i \circ \eta, \forall i = 1, 2$.*

To obtain sign-changing solutions, we should use cones of positive functions. Precisely, we define

$$\mathscr{P}_i := \{\mathbf{u} = (u_1, u_2) \in H \mid u_i \geq 0\}, \quad \mathscr{P} := \bigcup_{i=1}^{2} (\mathscr{P}_i \cup -\mathscr{P}_i). \qquad (3.24)$$

Clearly both u_1 and u_2 change sign if and only if $(u_1, u_2) \in H \setminus \mathscr{P}$. For $\delta > 0$ we define $\mathscr{P}_\delta := \{\mathbf{u} \in H \mid \text{dist}_4(\mathbf{u}, \mathscr{P}) < \delta\}$, where

$$\text{dist}_4(\mathbf{u}, \mathscr{P}) := \min\{\text{dist}_4(u_i, \mathscr{P}_i), \text{dist}_4(u_i, -\mathscr{P}_i), \quad i = 1, 2\}, \qquad (3.25)$$

$$\text{dist}_4(u_i, \pm\mathscr{P}_i) := \inf\{|u_i - v_i|_4 \mid v_i \in \pm\mathscr{P}_i\}.$$

Denote $u^\pm := \max\{0, \pm u\}$, then it is easy to check that $\text{dist}_4(u_i, \pm\mathscr{P}_i) = |u_i^\mp|_4$. The following lemma was proved in [26]. Since the proof is very simple, we give the details here for completeness.

Lemma 3.7 ([26, Lemma 2.6]) *Let $k_1, k_2 \geq 2$. Then for any $\delta < 2^{-1/4}$ and any $A \in \Gamma^{(k_1, k_2)}$ there holds $A \setminus \mathscr{P}_\delta \neq \emptyset$.*

Proof Fix any $A \in \Gamma^{(k_1,k_2)}$. Consider

$$f = (f_1, f_2) : A \to \mathbb{R}^{k_1-1} \times \mathbb{R}^{k_2-1}, \quad f_i(\mathbf{u}) = \left(\int_\Omega |u_i|^3 u_i \, dx, 0, \dots, 0 \right).$$

(3.26)

Clearly $f \in F_{(k_1,k_2)}(A)$, so there exists $\mathbf{u} \in A$ such that $f(\mathbf{u}) = 0$. Noting $\mathbf{u} \in A \subset \mathcal{M}$, we conclude that

$$\int_\Omega (u_i^+)^4 \, dx = \int_\Omega (u_i^-)^4 \, dx = 1/2, \quad \text{for} \quad i = 1, 2,$$

that is, $\mathrm{dist}_4(\mathbf{u}, \mathscr{P}) = 2^{-1/4}$, and so $\mathbf{u} \in A \setminus \mathscr{P}_\delta$ for every $\delta < 2^{-1/4}$.   □

Lemma 3.8 *There exists $A \in \Gamma^{(k+1,k+1)}$ such that $A \subset \mathcal{N}_b$ and $\sup_A J_\beta < d_k$.*

Proof Recalling W_{k+1} in (3.13), we define

$$A_1 = A_2 := \{ u \in W_{k+1} : |u|_4 = 1 \}.$$

Note that there exists an obvious odd homeomorphism from S^k to A_i. By Lemma 3.6-(i) one has $A := A_1 \times A_2 \in \Gamma^{(k+1,k+1)}$. On the other hand, we see from (3.13) that $A \subset \mathcal{N}_{\bar{b}} \subset \mathcal{N}_b$. Thus, Lemma 3.3 yields $\sup_A J_\beta < d_k$.   □

For every $k_1, k_2 \in [2, k+1]$ and $0 < \delta < 2^{-1/4}$, we define minimax values

$$c_{\beta,\delta}^{k_1,k_2} := \inf_{A \in \Gamma_\beta^{(k_1,k_2)}} \sup_{\mathbf{u} \in A \setminus \mathscr{P}_\delta} J_\beta(\mathbf{u}),$$

(3.27)

where

$$\Gamma_\beta^{(k_1,k_2)} := \left\{ A \in \Gamma^{(k_1,k_2)} \, \bigg| \, A \subset \mathcal{N}_b, \, \sup_A J_\beta < d_k \right\}.$$

(3.28)

Noting that $\Gamma_\beta^{(\tilde{k}_1,\tilde{k}_2)} \subset \Gamma_\beta^{(k_1,k_2)}$ for any $\tilde{k}_1 \geq k_1$ and $\tilde{k}_2 \geq k_2$, we see that Lemma 3.8 yields $\Gamma_\beta^{(k_1,k_2)} \neq \emptyset$ and so $c_{\beta,\delta}^{k_1,k_2}$ is well defined for any $k_1, k_2 \in [2, k+1]$. Moreover,

$$c_{\beta,\delta}^{k_1,k_2} < d_k \quad \text{for every} \quad \delta \in (0, 2^{-1/4}) \quad \text{and} \quad k_1, k_2 \in [2, k+1].$$

We will prove that $c_{\beta,\delta}^{k_1,k_2}$ is a sign-changing critical value of E_β for $\delta > 0$ sufficiently small. Define

$$\mathcal{N}_{b,\beta} := \{ \mathbf{u} \in \mathcal{N}_b \mid J_\beta(\mathbf{u}) < d_k \},$$

then Lemma 3.3 yields $\mathcal{N}_{\bar{b}} \subset \mathcal{N}_{b,\beta}$.

Lemma 3.9 *For any sufficiently small $\delta \in (0, 2^{-1/4})$, there holds*

$$\text{dist}_4(K(\mathbf{u}), \mathscr{P}) < \delta/2, \quad \forall \mathbf{u} \in \mathscr{N}_{b,\beta}, \ \text{dist}_4(\mathbf{u}, \mathscr{P}) < \delta.$$

Proof Assume by contradiction that there exist $\delta_n \to 0$ and $\mathbf{u}_n = (u_{n,1}, u_{n,2}) \in \mathscr{N}_{b,\beta}$ such that $\text{dist}_4(\mathbf{u}_n, \mathscr{P}) < \delta_n$ and $\text{dist}_4(K(\mathbf{u}_n), \mathscr{P}) \geq \delta_n/2$. Without loss of generality we may assume that $\text{dist}_4(\mathbf{u}_n, \mathscr{P}) = \text{dist}_4(u_{n,1}, \mathscr{P}_1)$. Write $K(\mathbf{u}_n) = \mathbf{w}_n = (w_{n,1}, w_{n,2})$ and $w_{n,i} = \alpha_{n,i}\tilde{w}_{n,i}$ as in Lemma 3.5. Then by the proof of Lemma 3.5, we see that $\alpha_{n,i}$ are all uniformly bounded. Combining this with (3.17) and (3.20), we deduce that

$$\begin{aligned}
\text{dist}_4(w_{n,1}, \mathscr{P}_1)|w_{n,1}^-|_4 = |w_{n,1}^-|_4^2 &\leq C\|w_{n,1}^-\|_{\lambda_1}^2 \\
&\leq C \int_\Omega \left(|\nabla w_{n,1}^-|^2 + \lambda_1(w_{n,1}^-)^2 - \beta t_2(\mathbf{u}_n) u_{n,2}^2(w_{n,1}^-)^2 \right) dx \\
&= -C\alpha_{n,1}\mu_1 t_1(\mathbf{u}_n) \int_\Omega u_{n,1}^3 w_{n,1}^- \, dx \\
&\leq C \int_\Omega (u_{n,1}^-)^3 w_{n,1}^- \, dx \leq C|u_{n,1}^-|_4^3 |w_{n,1}^-|_4 \\
&= C\text{dist}_4(u_{n,1}, \mathscr{P}_1)^3 |w_{n,1}^-|_4 \leq C\delta_n^3 |w_{n,1}^-|_4.
\end{aligned}$$

Consequently $\text{dist}_4(K(\mathbf{u}_n), \mathscr{P}) \leq \text{dist}_4(w_{n,1}, \mathscr{P}_1) \leq C\delta_n^3 < \delta_n/2$ for n sufficiently large, which is a contradiction. $\qquad\square$

Now let us define a map $V : \mathscr{N}_b^* \to H$ by $V(\mathbf{u}) := \mathbf{u} - K(\mathbf{u})$. We will prove that $(\sqrt{t_1(\mathbf{u})}u_1, \sqrt{t_2(\mathbf{u})}u_2)$ is a sign-changing solution of (3.1) if $\mathbf{u} = (u_1, u_2) \in \mathscr{N}_b \setminus \mathscr{P}$ satisfies $V(\mathbf{u}) = 0$. Here, any $\mathbf{u}$ satisfying $V(\mathbf{u}) = 0$ is actually a *fixed point* of K.

Lemma 3.10 *Let $\mathbf{u}_n = (u_{n,1}, u_{n,2}) \in \mathscr{N}_b$ be such that*

$$J_\beta(\mathbf{u}_n) \to c < d_k \quad and \quad V(\mathbf{u}_n) \to 0 \quad strongly \ in \ H.$$

Then up to a subsequence, there exists $\mathbf{u} \in \mathscr{N}_b$ such that $\mathbf{u}_n \to \mathbf{u}$ strongly in H and $V(\mathbf{u}) = 0$.

Proof By Lemma 3.5, up to a subsequence, we may assume that $\mathbf{u}_n \rightharpoonup \mathbf{u} = (u_1, u_2)$ weakly in H and $\mathbf{w}_n := K(\mathbf{u}_n) = (w_{n,1}, w_{n,2}) \to \mathbf{w} = (w_1, w_2)$ strongly in H. This, together with $V(\mathbf{u}_n) \to 0$, yields

$$\begin{aligned}
\int_\Omega \nabla u_{n,i} \nabla(u_{n,i} - u_i) \, dx &= \int_\Omega \nabla(w_{n,i} - w_i)\nabla(u_{n,i} - u_i) \, dx \\
&+ \int_\Omega \nabla w_i \nabla(u_{n,i} - u_i) \, dx + \int_\Omega \nabla(u_{n,i} - w_{n,i})\nabla(u_{n,i} - u_i) \, dx = o(1).
\end{aligned}$$

Then it is easy to see that $\mathbf{u}_n \to \mathbf{u}$ strongly in H. Consequently, $\mathbf{u} \in \overline{\mathcal{N}_b}$ and $V(\mathbf{u}) = \lim_{n\to\infty} V(\mathbf{u}_n) = 0$. Moreover, $J_\beta(\mathbf{u}) = c < d_k$ and so $\mathbf{u} \in \mathcal{N}_b$. $\qquad\square$

Lemma 3.11 *Recall $C_0 > 0$ in Lemma 3.2. Then*

$$J'_\beta(\mathbf{u})[V(\mathbf{u})] \geq \frac{C_0}{2} \|V(\mathbf{u})\|_H^2, \quad \text{for any} \quad \mathbf{u} \in \mathcal{N}_b.$$

Proof Fix any $\mathbf{u} = (u_1, u_2) \in \mathcal{N}_b$ and write $\mathbf{w} = K(\mathbf{u}) = (w_1, w_2)$ as above, then $V(\mathbf{u}) = (u_1 - w_1, u_2 - w_2)$. By (3.20) we have $\int_\Omega u_i^3(u_i - w_i)\, dx = 1 - 1 = 0$. Then we deduce from (3.9)–(3.10), (3.17) and (3.20) that

$$
\begin{aligned}
J'_\beta(\mathbf{u})[V(\mathbf{u})] &= \sum_{i=1}^2 t_i(\mathbf{u}) \int_\Omega \Big(\nabla u_i \nabla(u_i - w_i) + \lambda_i u_i(u_i - w_i) \\
&\quad - t_j(\mathbf{u})\beta u_i(u_i - w_i)u_j^2 \Big)\, dx \\
&= \sum_{i=1}^2 t_i(\mathbf{u}) \int_\Omega \Big(\nabla u_i \nabla(u_i - w_i) + \lambda_i u_i(u_i - w_i) \\
&\quad - t_j(\mathbf{u})\beta w_i(u_i - w_i)u_j^2 - t_j(\mathbf{u})\beta(u_i - w_i)^2 u_j^2 \Big)\, dx \\
&= \sum_{i=1}^2 t_i(\mathbf{u}) \int_\Omega \Big(\nabla u_i \nabla(u_i - w_i) + \lambda_i u_i(u_i - w_i) - \nabla w_i \nabla(u_i - w_i) \\
&\quad - \lambda_i w_i(u_i - w_i) + \alpha_i \mu_i t_i(\mathbf{u}) u_i^3(u_i - w_i) - t_j(\mathbf{u})\beta(u_i - w_i)^2 u_j^2 \Big)\, dx \\
&= \sum_{i=1}^2 t_i(\mathbf{u}) \int_\Omega \Big(|\nabla(u_i - w_i)|^2 + \lambda_i |u_i - w_i|^2 \\
&\quad - t_j(\mathbf{u})\beta(u_i - w_i)^2 u_j^2 \Big)\, dx \\
&\geq \sum_{i=1}^2 \frac{t_i(\mathbf{u})}{2} \|u_i - w_i\|_{\lambda_i}^2 \geq \frac{C_0}{2} \|V(\mathbf{u})\|_H^2.
\end{aligned}
$$

This completes the proof. $\qquad\square$

Lemma 3.12 *There exists a unique global solution $\eta = (\eta_1, \eta_2) : [0, \infty) \times \mathcal{N}_{b,\beta} \to H$ for the initial value problem*

$$\frac{d}{dt}\eta(t, \mathbf{u}) = -V(\eta(t, \mathbf{u})), \quad \eta(0, \mathbf{u}) = \mathbf{u} \in \mathcal{N}_{b,\beta}. \tag{3.29}$$

Moreover,

(i) $\eta(t, \mathbf{u}) \in \mathcal{N}_{b,\beta}$ *for any $t > 0$ and $\mathbf{u} \in \mathcal{N}_{b,\beta}$.*

(ii) $\eta(t, \sigma_i(\mathbf{u})) = \sigma_i(\eta(t, \mathbf{u}))$ *for any $t > 0$, $\mathbf{u} \in \mathcal{N}_{b,\beta}$ and $i = 1, 2$.*

(iii) *For every $\mathbf{u} \in \mathcal{N}_{b,\beta}$, the map $t \mapsto J_\beta(\eta(t, \mathbf{u}))$ is nonincreasing.*

(iv) *There exists $\delta_0 \in (0, 2^{-1/4})$ such that, for every $\delta < \delta_0$, there holds*

$$\eta(t, \mathbf{u}) \in \mathscr{P}_\delta \quad \text{whenever} \quad \mathbf{u} \in \mathcal{N}_{b,\beta} \cap \mathscr{P}_\delta \quad \text{and} \quad t > 0.$$

Proof Recalling Lemma 3.4, we have $V(\mathbf{u}) \in C^1(\mathcal{N}_b^*, H)$. Since $\mathcal{N}_{b,\beta} \subset \mathcal{N}_b^*$ and $\mathcal{N}_b^*$ is open, so (3.29) has a unique solution $\eta : [0, T_{\max}) \times \mathcal{N}_{b,\beta} \to H$, where $T_{\max} > 0$ is the maximal time such that $\eta(t, \mathbf{u}) \in \mathcal{N}_b^*$ for all $t \in [0, T_{\max})$ (remark that $V(\cdot)$ is defined only on $\mathcal{N}_b^*$). We should prove $T_{\max} = +\infty$ for any $\mathbf{u} \in \mathcal{N}_{b,\beta}$. Fixing any $\mathbf{u} = (u_1, u_2) \in \mathcal{N}_{b,\beta}$, we have

$$\frac{d}{dt} \int_\Omega \eta_i(t, \mathbf{u})^4 \, dx = -4 \int_\Omega \eta_i(t, \mathbf{u})^3 (\eta_i(t, \mathbf{u}) - K_i(\eta(t, \mathbf{u}))) \, dx$$

$$= 4 - 4 \int_\Omega \eta_i(t, \mathbf{u})^4 \, dx, \quad \forall 0 < t < T_{\max}.$$

Recalling $\int_\Omega \eta_i(0, \mathbf{u})^4 \, dx = \int_\Omega u_i^4 \, dx = 1$, we deduce that $\int_\Omega \eta_i(t, \mathbf{u})^4 \, dx \equiv 1$ for all $0 \leq t < T_{\max}$. So $\eta(t, \mathbf{u}) \in \mathcal{M}$, that is $\eta(t, \mathbf{u}) \in \mathcal{M} \cap \mathcal{N}_b^* = \mathcal{N}_b$ for all $t \in [0, T_{\max})$. Assume by contradiction that $T_{\max} < +\infty$, then $\eta(T_{\max}, \mathbf{u}) \in \partial \mathcal{N}_b$, and so $J_\beta(\eta(T_{\max}, \mathbf{u})) \geq d_k$. Since $\eta(t, \mathbf{u}) \in \mathcal{N}_b$ for any $t \in [0, T_{\max})$, we deduce from Lemma 3.11 that

$$J_\beta(\eta(T_{\max}, \mathbf{u})) = J_\beta(\mathbf{u}) - \int_0^{T_{\max}} J_\beta'(\eta(t, \mathbf{u}))[V(\eta(t, \mathbf{u}))] \, dt$$

$$\leq J_\beta(\mathbf{u}) - \frac{C_0}{2} \int_0^{T_{\max}} \|V(\eta(t, \mathbf{u}))\|_H^2 \, dt \leq J_\beta(\mathbf{u}) < d_k, \qquad (3.30)$$

a contradiction. So $T_{\max} = +\infty$. Then similarly as (3.30) we have $J_\beta(\eta(t, \mathbf{u})) \leq J_\beta(\mathbf{u}) < d_k$ for all $t > 0$, so $\eta(t, \mathbf{u}) \in \mathcal{N}_{b,\beta}$ and then (i), (iii) hold.

By (3.23) we have $V(\sigma_i(\mathbf{u})) = \sigma_i(V(\mathbf{u}))$. Then by the uniqueness of solutions of the initial value problem (3.29), it is easy to check that (ii) holds.

Finally, let $\delta_0 \in (0, 2^{-1/4})$ such that Lemma 3.9 holds for every $\delta < \delta_0$. For any $\mathbf{u} \in \mathcal{N}_{b,\beta}$ with $\text{dist}_4(\mathbf{u}, \mathscr{P}) = \delta < \delta_0$, since

$$\eta(t, \mathbf{u}) = \mathbf{u} + t \frac{d}{dt} \eta(0, \mathbf{u}) + o(t) = \mathbf{u} - t V(\mathbf{u}) + o(t) = (1 - t)\mathbf{u} + t K(\mathbf{u}) + o(t),$$

we deduce from Lemma 3.9 that

$$
\begin{aligned}
\text{dist}_4(\eta(t, \mathbf{u}), \mathscr{P}) &= \text{dist}_4((1-t)\mathbf{u} + tK(\mathbf{u}) + o(t), \mathscr{P}) \\
&\leq (1-t)\text{dist}_4(\mathbf{u}, \mathscr{P}) + t\text{dist}_4(K(\mathbf{u}), \mathscr{P}) + o(t) \\
&\leq (1-t)\delta + t\delta/2 + o(t) < \delta
\end{aligned}
$$

for $t > 0$ sufficiently small. Hence (iv) holds. $\square$

Now we are in the position to prove the existence of k sign-changing solutions.

Proof (Proof of Theorem 3.4: I Existence of k sign-changing solutions)

Step 1. Fix any $k_1, k_2 \in [2, k+1]$ and take any $\delta \in (0, \delta_0)$. We prove that (3.1) has a sign-changing solution $(\tilde{u}_1, \tilde{u}_2) \in H$ such that $E_\beta(\tilde{u}_1, \tilde{u}_2) = c_{\beta,\delta}^{k_1,k_2}$.

Write $c_{\beta,\delta}^{k_1,k_2}$ simply by c in this step. Recall that $c < d_k$. We claim that there exists a sequence $\{\mathbf{u}_n : n \geq 1\} \subset \mathscr{N}_{b,\beta}$ such that

$$
J_\beta(\mathbf{u}_n) \to c, \quad V(\mathbf{u}_n) \to 0 \text{ as } n \to \infty, \quad \text{and} \quad \text{dist}_4(\mathbf{u}_n, \mathscr{P}) \geq \delta, \quad \forall n \in \mathbb{N}.
$$
$$(3.31)$$

Suppose by contradiction that (3.31) does not hold. Then there exists small $\varepsilon \in (0, 1)$ such that

$$
\|V(\mathbf{u})\|_H^2 \geq \varepsilon, \quad \forall \mathbf{u} \in \mathscr{N}_{b,\beta}, \ |J_\beta(\mathbf{u}) - c| \leq 2\varepsilon, \quad \text{dist}_4(\mathbf{u}, \mathscr{P}) \geq \delta.
$$

Recalling the definition of c in (3.27), we can find $A \in \Gamma_\beta^{(k_1, k_2)}$ such that

$$
\sup_{A \backslash \mathscr{P}_\delta} J_\beta < c + \varepsilon.
$$

Since $\sup_A J_\beta < d_k$, we have $A \subset \mathscr{N}_{b,\beta}$ and so we can consider $B = \eta(4/C_0, A)$, where η is in Lemma 3.12 and C_0 is in Lemma 3.2. Lemma 3.12-(i) yields $B \subset \mathscr{N}_{b,\beta}$. By Lemma 3.6-(ii) and Lemma 3.12-(ii) we have $B \in \Gamma^{(k_1,k_2)}$. Again by Lemma 3.12-(iii), we have $\sup_B J_\beta \leq \sup_A J_\beta < d_k$, namely $B \in \Gamma_\beta^{(k_1,k_2)}$ and so $\sup_{B \backslash \mathscr{P}_\delta} J_\beta \geq c$. Consequently by Lemma 3.7 we can take $\mathbf{u} \in A$ such that $\eta(4/C_0, \mathbf{u}) \in B \backslash \mathscr{P}_\delta$ and

$$
c - \varepsilon \leq \sup_{B \backslash \mathscr{P}_\delta} J_\beta - \varepsilon < J_\beta(\eta(4/C_0, \mathbf{u})).
$$

Since $\eta(t, \mathbf{u}) \in \mathscr{N}_{b,\beta}$ for any $t \geq 0$, Lemma 3.12-(iv) yields $\eta(t, \mathbf{u}) \notin \mathscr{P}_\delta$ for any $t \in [0, 4/C_0]$. In particular, $\mathbf{u} \notin \mathscr{P}_\delta$ and so $J_\beta(\mathbf{u}) < c + \varepsilon$. Then for any $t \in [0, 4/C_0]$, we have

$$
c - \varepsilon < J_\beta(\eta(4/C_0, \mathbf{u})) \leq J_\beta(\eta(t, \mathbf{u})) \leq J_\beta(\mathbf{u}) < c + \varepsilon,
$$

which implies $\|V(\eta(t, \mathbf{u}))\|_H^2 \geq \varepsilon$ for every $t \in [0, 4/C_0]$. Consequently

$$\frac{d}{dt} J_\beta(\eta(t, \mathbf{u})) = -J_\beta'(\eta(t, \mathbf{u}))[V(\eta(t, \mathbf{u}))] \leq -\frac{C_0}{2}\|V(\eta(t, \mathbf{u}))\|_H^2 \leq -\frac{C_0}{2}\varepsilon$$

for every $t \in [0, 4/C_0]$. Hence,

$$c - \varepsilon < J_\beta(\eta(4/C_0, \mathbf{u})) \leq J_\beta(\mathbf{u}) - \int_0^{4/C_0} \frac{C_0}{2}\varepsilon \, dt < c + \varepsilon - 2\varepsilon = c - \varepsilon,$$

a contradiction. Therefore (3.31) holds. By Lemma 3.10, up to a subsequence, there exists $\mathbf{u} = (u_1, u_2) \in \mathcal{N}_{b,\beta}$ such that $\mathbf{u}_n \to \mathbf{u}$ strongly in H, where $\mathbf{u}$ satisfies $V(\mathbf{u}) = 0$ and $J_\beta(\mathbf{u}) = c = c_{\beta,\delta}^{k_1,k_2}$. Since $\mathrm{dist}_4(\mathbf{u}_n, \mathcal{P}) \geq \delta$, we see that $\mathrm{dist}_4(\mathbf{u}, \mathcal{P}) \geq \delta$, which implies that both u_1 and u_2 are sign-changing. Since $V(\mathbf{u}) = 0$, we have $\mathbf{u} = K(\mathbf{u})$. Combining this with (3.20)–(3.21), we see that $\mathbf{u}$ satisfies

$$\begin{cases} -\Delta u_1 + \lambda_1 u_1 = \alpha_1 \mu_1 t_1(\mathbf{u}) u_1^3 + \beta t_2(\mathbf{u}) u_2^2 u_1, \\ -\Delta u_2 + \lambda_2 u_2 = \alpha_2 \mu_2 t_2(\mathbf{u}) u_2^3 + \beta t_1(\mathbf{u}) u_1^2 u_2. \end{cases} \tag{3.32}$$

Recall that $|u_i|_4 = 1$ and $t_i(\mathbf{u})$ satisfies (3.5). Multiplying (3.32) by u_i and integrating over Ω, we easily obtain that $\alpha_1 = \alpha_2 = 1$. Again by (3.32), we conclude that $(\tilde{u}_1, \tilde{u}_2) := (\sqrt{t_1(\mathbf{u})}u_1, \sqrt{t_2(\mathbf{u})}u_2)$ is a sign-changing solution of the original problem (3.1). Moreover, (3.6) and (3.11) yield $E_\beta(\tilde{u}_1, \tilde{u}_2) = J_\beta(u_1, u_2) = c_{\beta,\delta}^{k_1,k_2}$.

Step 2. We prove that (3.1) has at least k sign-changing solutions.

Assume by contradiction that (3.1) has at most $k - 1$ sign-changing solutions. Fix any $k_2 \in [2, k + 1]$ and $\delta \in (0, \delta_0)$. Since $\Gamma_\beta^{(k_1+1,k_2)} \subset \Gamma_\beta^{(k_1,k_2)}$, we have

$$c_{\beta,\delta}^{2,k_2} \leq c_{\beta,\delta}^{3,k_2} \leq \cdots \leq c_{\beta,\delta}^{k,k_2} \leq c_{\beta,\delta}^{k+1,k_2} < d_k. \tag{3.33}$$

Since $c_{\beta,\delta}^{k_1,k_2}$ is a sign-changing critical value of E_β for each $k_1 \in [2, k + 1]$ (that is, E_β has a sign-changing critical point $\mathbf{u}$ with $E_\beta(\mathbf{u}) = c_{\beta,\delta}^{k_1,k_2}$), by (3.33) and our assumption that (3.1) has at most $k - 1$ sign-changing solutions, there exists some $2 \leq N_1 \leq k$ such that

$$c_{\beta,\delta}^{N_1,k_2} = c_{\beta,\delta}^{N_1+1,k_2} =: \bar{c} < d_k. \tag{3.34}$$

Define

$$\mathscr{K} := \{\mathbf{u} \in \mathcal{N}_b \mid \mathbf{u} \text{ is sign-changing}, \quad J_\beta(\mathbf{u}) = \bar{c}, \ V(\mathbf{u}) = 0\}. \tag{3.35}$$

Then $\mathcal{K}$ is finite. By (3.23) one has that $\sigma_i(\mathbf{u}) \in \mathcal{K}$ if $\mathbf{u} \in \mathcal{K}$, namely $\mathcal{K} \subset \mathscr{F}$. Hence there exist $k_0 \le k - 1$ and $\{\mathbf{u}_m : 1 \le m \le k_0\} \subset \mathcal{K}$ such that

$$\mathcal{K} = \{\mathbf{u}_m, \sigma_1(\mathbf{u}_m), \sigma_2(\mathbf{u}_m), -\mathbf{u}_m \mid 1 \le m \le k_0\}.$$

Consequently there exist open neighborhoods $O_{\mathbf{u}_m}$ of $\mathbf{u}_m$ in H, such that any two of $\overline{O_{\mathbf{u}_m}}$, $\sigma_1(\overline{O_{\mathbf{u}_m}})$, $\sigma_2(\overline{O_{\mathbf{u}_m}})$ and $-\overline{O_{\mathbf{u}_m}}$, where $1 \le m \le k_0$, are disjointed and

$$\mathcal{K} \subset O := \bigcup_{m=1}^{k_0} O_{\mathbf{u}_m} \cup \sigma_1(O_{\mathbf{u}_m}) \cup \sigma_2(O_{\mathbf{u}_m}) \cup -O_{\mathbf{u}_m}.$$

Define a continuous map $\tilde{f} : \overline{O} \to \mathbb{R} \setminus \{0\}$ by

$$\tilde{f}(\mathbf{u}) := \begin{cases} 1, & \text{if } \mathbf{u} \in \bigcup_{m=1}^{k_0} \overline{O_{\mathbf{u}_m}} \cup \sigma_2(\overline{O_{\mathbf{u}_m}}), \\ -1, & \text{if } \mathbf{u} \in \bigcup_{m=1}^{k_0} \sigma_1(\overline{O_{\mathbf{u}_m}}) \cup -\overline{O_{\mathbf{u}_m}}. \end{cases}$$

Clearly $\tilde{f}(\sigma_1(\mathbf{u})) = -\tilde{f}(\mathbf{u})$ and $\tilde{f}(\sigma_2(\mathbf{u})) = \tilde{f}(\mathbf{u})$. By Tietze's extension theorem, there exists $f \in C(H, \mathbb{R})$ such that $f|_O \equiv \tilde{f}$. Define

$$F(\mathbf{u}) := \frac{f(\mathbf{u}) + f(\sigma_2(\mathbf{u})) - f(\sigma_1(\mathbf{u})) - f(-\mathbf{u})}{4},$$

then $F|_O \equiv \tilde{f}$, $F(\sigma_1(\mathbf{u})) = -F(\mathbf{u})$ and $F(\sigma_2(\mathbf{u})) = F(\mathbf{u})$. Define

$$\mathcal{K}_\tau := \left\{ \mathbf{u} \in \mathcal{N}_b \mid \inf_{\mathbf{v} \in \mathcal{K}} \|\mathbf{u} - \mathbf{v}\|_H < \tau \right\}.$$

Then we can take small $\tau > 0$ such that $\mathcal{K}_{2\tau} \subset O$. Recalling $V(\mathbf{u}) = 0$ in $\mathcal{K}$ and $\mathcal{K}$ finite, we see that there exists $\tilde{C} > 0$ such that

$$\|V(\mathbf{u})\|_H \le \tilde{C}, \quad \forall \mathbf{u} \in \overline{\mathcal{K}_{2\tau}}. \tag{3.36}$$

Remark that for any $\mathbf{u} \in \mathcal{K}_{2\tau}$, we have $F(\mathbf{u}) = \tilde{f}(\mathbf{u}) \ne 0$, namely $F(\mathcal{K}_{2\tau}) \subset \mathbb{R} \setminus \{0\}$. On the other hand, by (3.35) and Lemma 3.10 there exists small $\varepsilon \in (0, (d_k - \bar{c})/2)$ such that

$$\|V(\mathbf{u})\|_H^2 \ge \varepsilon, \quad \forall u \in \mathcal{N}_b \setminus (\mathcal{K}_\tau \cup \mathscr{P}_\delta) \text{ satisfying } |J_\beta(\mathbf{u}) - \bar{c}| \le 2\varepsilon. \tag{3.37}$$

Recalling C_0 in Lemma 3.2 and $\tilde{C}$ in (3.36), we let

$$\alpha := \frac{1}{2} \min \left\{ 1, \frac{\tau C_0}{2\tilde{C}} \right\}. \tag{3.38}$$

By (3.27)–(3.28) and (3.34) we can find $A \in \Gamma_\beta^{(N_1+1,k_2)}$ such that

$$\sup_{A \setminus \mathscr{P}_\delta} J_\beta < c_{\beta,\delta}^{N_1+1,k_2} + \alpha\varepsilon/2 = \bar{c} + \alpha\varepsilon/2. \tag{3.39}$$

Let $B := A \setminus \mathscr{K}_{2\tau}$, then it is easy to check that $B \subset \mathscr{F}$. We claim that $\gamma(B) \geq (N_1, k_2)$. If not, there exists $\tilde{g} \in F_{(N_1,k_2)}(B)$ such that $\tilde{g}(\mathbf{u}) \neq 0$ for any $\mathbf{u} \in B$. By Tietze's extension theorem again, there exists $\bar{g} = (\bar{g}_1, \bar{g}_2) \in C(H, \mathbb{R}^{N_1-1} \times \mathbb{R}^{k_2-1})$ such that $\bar{g}|_B \equiv \tilde{g}$. Define $g = (g_1, g_2) \in C(H, \mathbb{R}^{N_1-1} \times \mathbb{R}^{k_2-1})$ by

$$g_1(\mathbf{u}) := \frac{\bar{g}_1(\mathbf{u}) + \bar{g}_1(\sigma_2(\mathbf{u})) - \bar{g}_1(\sigma_1(\mathbf{u})) - \bar{g}_1(-\mathbf{u})}{4},$$

$$g_2(\mathbf{u}) := \frac{\bar{g}_2(\mathbf{u}) + \bar{g}_2(\sigma_1(\mathbf{u})) - \bar{g}_2(\sigma_2(\mathbf{u})) - \bar{g}_2(-\mathbf{u})}{4},$$

then $g|_B \equiv \tilde{g}$, $g_i(\sigma_i(\mathbf{u})) = -g_i(\mathbf{u})$ and $g_i(\sigma_j(\mathbf{u})) = g_i(\mathbf{u})$ for $j \neq i$. Finally we define $G = (G_1, G_2) \in C(A, \mathbb{R}^{N_1+1-1} \times \mathbb{R}^{k_2-1})$ by

$$G_1(\mathbf{u}) := (F(\mathbf{u}), \quad g_1(\mathbf{u})) \in \mathbb{R}^{N_1+1-1}, \quad G_2(\mathbf{u}) := g_2(\mathbf{u}) \in \mathbb{R}^{k_2-1}.$$

By our constructions of F and g, we have $G \in F_{(N_1+1,k_2)}(A)$. Since $\gamma(A) \geq (N_1 + 1, k_2)$, so $G(\mathbf{u}) = 0$ for some $\mathbf{u} \in A$. If $\mathbf{u} \in \mathscr{K}_{2\tau}$, then $F(\mathbf{u}) \neq 0$, a contradiction. So $\mathbf{u} \in A \setminus \mathscr{K}_{2\tau} = B$, and then $g(\mathbf{u}) = \tilde{g}(\mathbf{u}) \neq 0$, also a contradiction. Hence $\gamma(B) \geq (N_1, k_2)$. Note that $B \subset A \subset \mathscr{N}_b$ and $\sup_B J_\beta \leq \sup_A J_\beta < d_k$, we see that $B \subset \mathscr{N}_{b,\beta}$ and $B \in \Gamma_\beta^{(N_1,k_2)}$. Then we can consider $D := \eta(\tau/(2\tilde{C}), B)$, where η is in Lemma 3.12 and $\tilde{C}$ is in (3.36). By Lemma 3.6-(ii) and Lemma 3.12, we have $D \subset \mathscr{N}_{b,\beta}$, $D \in \Gamma^{(N_1,k_2)}$ and $\sup_D J_\beta \leq \sup_B J_\beta < d_k$, namely $D \in \Gamma_\beta^{(N_1,k_2)}$. Consequently we see from (3.27)–(3.28) and (3.34) that

$$\sup_{D \setminus \mathscr{P}_\delta} J_\beta \geq c_{\beta,\delta}^{N_1,k_2} = \bar{c}.$$

By Lemma 3.7 we can take $\mathbf{u} \in B$ such that $\eta(\tau/(2\tilde{C}), \mathbf{u}) \in D \setminus \mathscr{P}_\delta$ and

$$\bar{c} - \alpha\varepsilon/2 \leq \sup_{D \setminus \mathscr{P}_\delta} J_\beta - \alpha\varepsilon/2 < J_\beta(\eta(\tau/(2\tilde{C}), \mathbf{u})).$$

Since $\eta(t, \mathbf{u}) \in \mathscr{N}_{b,\beta}$ for any $t \geq 0$, Lemma 3.12-(iv) yields $\eta(t, \mathbf{u}) \notin \mathscr{P}_\delta$ for any $t \in [0, \tau/(2\tilde{C})]$. In particular, $\mathbf{u} \notin \mathscr{P}_\delta$ and so (3.39) yields $J_\beta(\mathbf{u}) < \bar{c} + \alpha\varepsilon/2$. Then for any $t \in [0, \tau/(2\tilde{C})]$, we have

$$\bar{c} - \alpha\varepsilon/2 < J_\beta(\eta(\tau/(2\tilde{C}), \mathbf{u})) \leq J_\beta(\eta(t, \mathbf{u})) \leq J_\beta(\mathbf{u}) < \bar{c} + \alpha\varepsilon/2.$$

Recall that $\mathbf{u} \in B = A \setminus \mathscr{K}_{2\tau}$. If there exists $T \in (0, \tau/(2\tilde{C}))$ such that $\eta(T, \mathbf{u}) \in \mathscr{K}_\tau$, then there exist $0 \leq t_1 < t_2 \leq T$ such that $\eta(t_1, \mathbf{u}) \in \partial\mathscr{K}_{2\tau}$, $\eta(t_2, \mathbf{u}) \in \partial\mathscr{K}_\tau$ and

$\eta(t, \mathbf{u}) \in \mathscr{K}_{2\tau} \setminus \mathscr{K}_{\tau}$ for any $t \in (t_1, t_2)$. So we see from (3.36) that

$$\tau \le \|\eta(t_1, \mathbf{u}) - \eta(t_2, \mathbf{u})\|_H = \left\| \int_{t_1}^{t_2} V(\eta(t, \mathbf{u})) \, dt \right\|_H \le 2\widetilde{C}(t_2 - t_1),$$

that is, $\tau/(2\widetilde{C}) \le t_2 - t_1 \le T$, a contradiction. Hence $\eta(t, \mathbf{u}) \notin \mathscr{K}_{\tau}$ for any $t \in (0, \tau/(2\widetilde{C}))$. Then as Step 1, we deduce from (3.37) and (3.38) that

$$\bar{c} - \frac{\alpha \varepsilon}{2} < J_\beta(\eta(\tau/(2\widetilde{C}), \mathbf{u})) \le J_\beta(\mathbf{u}) - \int_0^{\frac{\tau}{2\widetilde{C}}} \frac{C_0}{2} \varepsilon \, dt < \bar{c} + \frac{\alpha \varepsilon}{2} - \alpha \varepsilon = \bar{c} - \frac{\alpha \varepsilon}{2},$$

which yields a contradiction. Therefore, system (3.1) has at least k sign-changing solutions for any $\beta \in (0, \beta_k)$. $\qquad \square$

Remark 3.3 If $A \in \Gamma^{(k_1, k_2)} \setminus \Gamma_\beta^{(k_1, k_2)}$, we cannot consider the set $\eta(4/C_0, A)$ in the proof of Theorem 3.4, because $\eta(t, \cdot)$ can not be defined on the whole $\mathscr{M}$ for any $t > 0$ and so $\eta(4/C_0, A)$ is not well defined. Hence we can not replace $\Gamma_\beta^{(k_1, k_2)}$ by $\Gamma^{(k_1, k_2)}$ in the definition of $c_{\beta, \delta}^{k_1, k_2}$.

Before we turn to prove the existence of k semi-nodal solutions, we give the proof of Theorem 3.5 first.

Proof (Proof of Theorem 3.5) Let $k = 1$ in Theorem 3.4. Then there exists $\beta_1 > 0$ such that, for any $\beta \in (0, \beta_1)$, (3.1) has a sign-changing solution $(u_{\beta, 1}, v_{\beta, 1})$ with $E_\beta(u_{\beta, 1}, v_{\beta, 1}) = c_{\beta, \delta}^{2,2} < d_1$. Recalling $\mathscr{C}$ in (3.15), we define

$$\beta_1' := \min \left\{ \mathscr{C}^2/(4d_1), \beta_1 \right\}. \tag{3.40}$$

Fix any $\beta \in (0, \beta_1')$ and define

$$c_\beta := \inf_{\mathbf{u} \in \mathscr{K}_\beta} E_\beta(\mathbf{u}); \quad \mathscr{K}_\beta := \{\mathbf{u} : \mathbf{u} \text{ is a sign-changing solution of (3.1)}\}.$$

Then $\mathscr{K}_\beta \ne \emptyset$ and $c_\beta < d_1$. Let $\mathbf{u}_n = (u_{n,1}, u_{n,2}) \in \mathscr{K}_\beta$ be a minimizing sequence of c_β with $E_\beta(\mathbf{u}_n) < d_1$ for all $n \ge 1$. Then $\|u_{n,1}\|_{\lambda_1}^2 + \|u_{n,2}\|_{\lambda_2}^2 < 4d_1$. Up to a subsequence, we may assume that $\mathbf{u}_n \rightharpoonup \mathbf{u} = (u_1, u_2)$ weakly in H and strongly in $L^4(\Omega) \times L^4(\Omega)$. Since $E_\beta'(\mathbf{u}_n) = 0$, it is standard to prove that $\mathbf{u}_n \to \mathbf{u} = (u_1, u_2)$ strongly in H, $E_\beta'(\mathbf{u}) = 0$ and $E_\beta(\mathbf{u}) = c_\beta$. On the other hand, we deduce from $E_\beta'(\mathbf{u}_n)(u_{n,1}^\pm, 0) = 0$ and $E_\beta'(\mathbf{u}_n)(0, u_{n,2}^\pm) = 0$ that

$$\mathscr{C} |u_{n,i}^\pm|_4^2 \le \|u_{n,i}^\pm\|_{\lambda_i}^2 = \mu_i |u_{n,i}^\pm|_4^4 + \beta \int_\Omega |u_{n,i}^\pm|^2 u_{n,j}^2 \, dx \le \mu_i |u_{n,i}^\pm|_4^4 + \beta |u_{n,i}^\pm|_4^2 |u_{n,j}|_4^2$$

$$\le \mu_i |u_{n,i}^\pm|_4^4 + \frac{\beta}{\mathscr{C}} |u_{n,i}^\pm|_4^2 \|u_{n,j}\|_{\lambda_j}^2 < \mu_i |u_{n,i}^\pm|_4^4 + \frac{4d_1 \beta}{\mathscr{C}} |u_{n,i}^\pm|_4^2,$$

which implies that $|u_{n,i}^\pm|_4 \geq C > 0$ for all $n \geq 1$ and $i = 1, 2$, where C is a constant independent of n and i. Hence $|u_i^\pm|_4 \geq C$ and then $\mathbf{u}$ is a least energy sign-changing solution of (3.1).   □

Remark 3.4 The proof of Theorem 3.5 here is different from Theorem 3.2 in [26]. For Theorem 3.2, since $\beta < 0$, we did not need to use minimizing arguments. Instead, we proved in [26] that the sign-changing least energy c_β is actually the same as the minimax value $c_{\beta,\delta}^{2,2}$. That is, the sign-changing solution (u_1, u_2) corresponding to the sign-changing critical value $c_{\beta,\delta}^{2,2}$ is actually a least energy sign-changing solution. However, for the attractive case $\beta > 0$ we consider in this chapter, the above idea does not apply. This provides a new evidence that the attractive case $\beta > 0$ is different from the repulsive case $\beta < 0$.

3.3 Semi-nodal Solutions

In this section, we continue the proof of Theorem 3.4 to show the existence of semi-nodal solutions. The following arguments are similar to those in Sect. 3.2 with some modifications. Here, although some definitions are slight different from those in Sect. 3.2, we will use the same notations as in Sect. 3.2 for convenience. To obtain semi-nodal solutions (u_1, u_2) such that u_1 changes sign and u_2 is positive, we consider the following modified functional

$$\widetilde{E}_\beta(u_1, u_2) := \frac{1}{2}\left(\|u_1\|_{\lambda_1}^2 + \|u_2\|_{\lambda_2}^2\right) - \frac{1}{4}\left(\mu_1|u_1|_4^4 + \mu_2|u_2^+|_4^4\right) - \frac{\beta}{2}\int_\Omega u_1^2 u_2^2 \, dx,$$

and modify the definition of $\widetilde{H}$ by $\widetilde{H} := \{(u_1, u_2) \in H : u_1 \neq 0, u_2^+ \neq 0\}$. Then by similar proofs as in Sect. 3.2, we have the following lemma.

Lemma 3.13 *For any* $\mathbf{u} = (u_1, u_2) \in \widetilde{H}$, *if*

$$\begin{cases} \mu_2|u_2^+|_4^4\|u_1\|_{\lambda_1}^2 - \beta\|u_2\|_{\lambda_2}^2\int_\Omega u_1^2 u_2^2 \, dx > 0, \\ \mu_1|u_1|_4^4\|u_2\|_{\lambda_2}^2 - \beta\|u_1\|_{\lambda_1}^2\int_\Omega u_1^2 u_2^2 \, dx > 0, \end{cases} \tag{3.41}$$

then system

$$\begin{cases} \|u_1\|_{\lambda_1}^2 = t_1\mu_1|u_1|_4^4 + t_2\beta\int_\Omega u_1^2 u_2^2 \, dx \\ \|u_2\|_{\lambda_2}^2 = t_2\mu_2|u_2^+|_4^4 + t_1\beta\int_\Omega u_1^2 u_2^2 \, dx \end{cases} \tag{3.42}$$

has a unique solution

$$
\begin{cases}
t_1(\mathbf{u}) = \dfrac{\mu_2 |u_2^+|_4^4 \|u_1\|_{\lambda_1}^2 - \beta \|u_2\|_{\lambda_2}^2 \int_\Omega u_1^2 u_2^2 \, dx}{\mu_1 \mu_2 |u_1|_4^4 |u_2^+|_4^4 - \beta^2 (\int_\Omega u_1^2 u_2^2 \, dx)^2} > 0 \\[3mm]
t_2(\mathbf{u}) = \dfrac{\mu_1 |u_1|_4^4 \|u_2\|_{\lambda_2}^2 - \beta \|u_1\|_{\lambda_1}^2 \int_\Omega u_1^2 u_2^2 \, dx}{\mu_1 \mu_2 |u_1|_4^4 |u_2^+|_4^4 - \beta^2 (\int_\Omega u_1^2 u_2^2)^2 \, dx} > 0.
\end{cases}
\tag{3.43}
$$

Moreover,

$$
\sup_{t_1, t_2 \ge 0} \widetilde{E}_\beta \left(\sqrt{t_1} u_1, \sqrt{t_2} u_2 \right) = \widetilde{E}_\beta \left(\sqrt{t_1(\mathbf{u})} u_1, \sqrt{t_2(\mathbf{u})} u_2 \right)
$$

$$
= \frac{1}{4} \frac{\mu_2 |u_2^+|_4^4 \|u_1\|_{\lambda_1}^4 - 2\beta \|u_1\|_{\lambda_1}^2 \|u_2\|_{\lambda_2}^2 \int_\Omega u_1^2 u_2^2 \, dx + \mu_1 |u_1|_4^4 \|u_2\|_{\lambda_2}^4}{\mu_1 \mu_2 |u_1|_4^4 |u_2^+|_4^4 - \beta^2 (\int_\Omega u_1^2 u_2^2 \, dx)^2}
\tag{3.44}
$$

and $(t_1(\mathbf{u}), t_2(\mathbf{u}))$ is the unique maximum point of $\widetilde{E}_\beta(\sqrt{t_1} u_1, \sqrt{t_2} u_2)$.

Now we modify the definitions of $\mathscr{M}^*$, $\mathscr{M}_\beta^*$, $\mathscr{M}_\beta^{**}$, $\mathscr{M}$, and $\mathscr{M}_\beta$ by

$$
\mathscr{M}^* := \left\{ \mathbf{u} \in H \mid 1/2 < |u_1|_4^4 < 2, \ 1/2 < |u_2^+|_4^4 < 2 \right\},
\tag{3.45}
$$

$$
\mathscr{M}_\beta^* := \left\{ \mathbf{u} \in \mathscr{M}^* \mid \mathbf{u} \text{ satisfies } (3.41) \right\},
$$

$$
\mathscr{M} := \left\{ \mathbf{u} \in H \mid |u_1|_4 = 1, \ |u_2^+|_4 = 1 \right\}, \quad \mathscr{M}_\beta := \mathscr{M} \cap \mathscr{M}_\beta^*,
$$

$$
\mathscr{M}_\beta^{**} := \left\{ \mathbf{u} \in \mathscr{M}^* \ \middle| \ \begin{array}{l} \mu_2 \|u_1\|_{\lambda_1}^2 - \beta \|u_2\|_{\lambda_2}^2 \int_\Omega u_1^2 u_2^2 \, dx > 0 \\[2mm] \mu_1 \|u_2\|_{\lambda_2}^2 - \beta \|u_1\|_{\lambda_1}^2 \int_\Omega u_1^2 u_2^2 \, dx > 0 \end{array} \right\},
\tag{3.46}
$$

and define a new functional $J_\beta : \mathscr{M}_\beta^{**} \to (0, +\infty)$ as in Sect. 3.2 by

$$
J_\beta(\mathbf{u}) := \frac{1}{4} \frac{\mu_2 \|u_1\|_{\lambda_1}^4 - 2\beta \|u_1\|_{\lambda_1}^2 \|u_2\|_{\lambda_2}^2 \int_\Omega u_1^2 u_2^2 \, dx + \mu_1 \|u_2\|_{\lambda_2}^4}{\mu_1 \mu_2 - \beta^2 (\int_\Omega u_1^2 u_2^2 \, dx)^2}.
$$

Then $J_\beta \in C^1(\mathscr{M}_\beta^{**}, (0, +\infty))$ and (3.9)–(3.10) hold for any $\mathbf{u} \in \mathscr{M}_\beta$ and $\varphi, \psi \in H_0^1(\Omega)$. Moreover, Lemma 3.13 yields

$$
J_\beta(u_1, u_2) = \sup_{t_1, t_2 \ge 0} \widetilde{E}_\beta \left(\sqrt{t_1} u_1, \sqrt{t_2} u_2 \right), \quad \forall (u_1, u_2) \in \mathscr{M}_\beta.
\tag{3.47}
$$

Under this new definitions (3.45)–(3.46), we define $\mathscr{N}_b^*$ and $\mathscr{N}_b$ as in (3.12)–(3.14). Since $|u_2|_4^2 \leq \mathscr{C}^{-1} \|u_2\|_{\lambda_2}^2 \leq b/\mathscr{C}$ for all $\mathbf{u} \in \mathscr{N}_b^*$, by trivial modifications it is easy to check that Lemmas 3.2 and 3.3 also hold here. Moreover, we may assume that (3.17) also holds here for any $\beta \in (0, \beta_k)$.

Now we fix any $\beta \in (0, \beta_k)$. For any $\mathbf{u} = (u_1, u_2) \in \mathscr{N}_b^*$, let $\tilde{w}_i \in H_0^1(\Omega)$, $i = 1, 2$, be the unique solutions of the following linear problem

$$\begin{cases} -\Delta \tilde{w}_1 + \lambda_1 \tilde{w}_1 - \beta t_2(\mathbf{u}) u_2^2 \tilde{w}_1 = \mu_1 t_1(\mathbf{u}) u_1^3, & \tilde{w}_1 \in H_0^1(\Omega), \\ -\Delta \tilde{w}_2 + \lambda_2 \tilde{w}_2 - \beta t_1(\mathbf{u}) u_1^2 \tilde{w}_2 = \mu_2 t_2(\mathbf{u})(u_2^+)^3, & \tilde{w}_2 \in H_0^1(\Omega). \end{cases} \tag{3.48}$$

As in Sect. 3.2, we define

$$w_i = \alpha_i \tilde{w}_i, \quad \text{where} \quad \alpha_1 = \frac{1}{\int_\Omega u_1^3 \tilde{w}_1 \, dx} > 0, \quad \alpha_2 = \frac{1}{\int_\Omega (u_2^+)^3 \tilde{w}_2 \, dx} > 0. \tag{3.49}$$

Then (w_1, w_2) is the unique solution of the problem

$$\begin{cases} -\Delta w_1 + \lambda_1 w_1 - \beta t_2(\mathbf{u}) u_2^2 w_1 = \alpha_1 \mu_1 t_1(\mathbf{u}) u_1^3, & w_1 \in H_0^1(\Omega), \\ -\Delta w_2 + \lambda_2 w_2 - \beta t_1(\mathbf{u}) u_1^2 w_2 = \alpha_2 \mu_2 t_2(\mathbf{u})(u_2^+)^3, & w_2 \in H_0^1(\Omega), \\ \int_\Omega u_1^3 w_1 \, dx = 1, \quad \int_\Omega (u_2^+)^3 w_2 \, dx = 1. \end{cases} \tag{3.50}$$

As in Sect. 3.2, the operator $K = (K_1, K_2) : \mathscr{N}_b^* \to H$ is defined as $K(\mathbf{u}) := \mathbf{w} = (w_1, w_2)$, and a similar argument as Lemma 3.4 yields $K \in C^1(\mathscr{N}_b^*, H)$. Since $u_n \to u$ in $L^4(\Omega)$ implies $u_n^+ \to u^+$ in $L^4(\Omega)$, we see that Lemma 3.5 also holds for this new K defined here. Clearly

$$K(\sigma_1(\mathbf{u})) = \sigma_1(K(\mathbf{u})). \tag{3.51}$$

Remark that (3.51) only holds for σ_1 and in the sequel we only use σ_1. Consider

$$\mathscr{F} = \{A \subset \mathscr{M} : A \text{ is closed and } \sigma_1(\mathbf{u}) \in A \,\forall\, \mathbf{u} \in A\},$$

and, for each $A \in \mathscr{F}$ and $k_1 \geq 2$, the class of functions

$$F_{(k_1, 1)}(A) = \left\{ f : A \to \mathbb{R}^{k_1 - 1} : f \text{ continuous and } f(\sigma_1(\mathbf{u})) = -f(\mathbf{u}) \right\}.$$

Definition 3.3 (*Modified vector genus, slightly different from Definition* 3.2) Let $A \in \mathscr{F}$ and take any $k_1 \in \mathbb{N}$ with $k_1 \geq 2$. We say that $\gamma(A) \geq (k_1, 1)$ if for every $f \in F_{(k_1, 1)}(A)$ there exists $\mathbf{u} \in A$ such that $f(\mathbf{u}) = 0$. We denote

$$\Gamma^{(k_1, 1)} := \{A \in \mathscr{F} : \gamma(A) \geq (k_1, 1)\}.$$

The following lemma is the counterpart of Lemma 3.6 for the modified vector genus.

Lemma 3.14 ([26, Lemma 4.2]) *With the previous notations, the following properties hold.*

(i) *Take $A := A_1 \times A_2 \subset \mathcal{M}$ and let $\eta : S^{k_1-1} \to A_1$ be a homeomorphism such that $\eta(-x) = -\eta(x)$ for every $x \in S^{k_1-1}$. Then $A \in \Gamma^{(k_1,1)}$.*
(ii) *We have $\overline{\eta(A)} \in \Gamma^{(k_1,1)}$ whenever $A \in \Gamma^{(k_1,1)}$ and a continuous map $\eta : A \to \mathcal{M}$ is such that $\eta \circ \sigma_1 = \sigma_1 \circ \eta$.*

Proof The proof is very simple, and we give it here for completeness. Noting that the conclusion (ii) is trivial, we only prove (i). Fix any $f \in F_{(k_1,1)}(A)$ and take any $u_2 \in A_2$. Define $\varphi : S^{k_1-1} \to \mathbb{R}^{k_1-1}$ by $\varphi(x) := f(\eta(x), u_2)$. Then φ is continuous and $\varphi(-x) = -\varphi(x)$. So by the classical Borsuk–Ulam theorem, there exists $x_0 \in S^{k_1-1}$ such that $\varphi(x_0) = 0$, namely $f(\eta(x_0), u_2) = 0$. Therefore, $\gamma(A) \geq (k_1, 1)$ and $A \in \Gamma^{(k_1,1)}$. $\qquad\square$

Now we modify the definitions of $\mathscr{P}$ and $\mathrm{dist}_4(\mathbf{u}, \mathscr{P})$ in (3.24)–(3.25) by

$$\mathscr{P} := \mathscr{P}_1 \cup -\mathscr{P}_1, \quad \mathrm{dist}_4(\mathbf{u}, \mathscr{P}) := \min\left\{\mathrm{dist}_4(u_1, \mathscr{P}_1), \mathrm{dist}_4(u_1, -\mathscr{P}_1)\right\}. \tag{3.52}$$

Under this new definition, we see that u_1 changes sign if and only if $\mathrm{dist}_4(\mathbf{u}, \mathscr{P}) > 0$.

Lemma 3.15 ([26, Lemma 4.3]) *Let $k_1 \geq 2$. Then for any $\delta < 2^{-1/4}$ and any $A \in \Gamma^{(k_1,1)}$ there holds $A \setminus \mathscr{P}_\delta \neq \emptyset$.*

The proof of Lemma 3.15 is a trivial modification from that of Lemma 3.7, so we omit the details here.

Lemma 3.16 *There exists $A \in \Gamma^{(k+1,1)}$ such that $A \subset \mathcal{N}_b$ and $\sup_A J_\beta < d_k$.*

Proof Recalling that $\varphi_0 \in W_{k+1}$ is positive, we define

$$A_1 := \left\{u \in W_{k+1} : |u|_4 = 1\right\}, \quad A_2 := \{C\varphi_0 : C = 1/|\varphi_0|_4\}.$$

Then by Lemma 3.14-(i) one has $A := A_1 \times A_2 \in \Gamma^{(k+1,1)}$. The rest of the proof is the same as Lemma 3.8. $\qquad\square$

For every $k_1 \in [2, k+1]$ and $0 < \delta < 2^{-1/4}$, we define minimax values

$$c_{\beta,\delta}^{k_1,1} := \inf_{A \in \Gamma_\beta^{(k_1,1)}} \sup_{\mathbf{u} \in A \setminus \mathscr{P}_\delta} J_\beta(\mathbf{u}),$$

where the definition of $\Gamma_\beta^{(k_1,1)}$ is the same as (3.28). Then Lemma 3.16 yields $\Gamma_\beta^{(k_1,1)} \neq \emptyset$ and so $c_{\beta,\delta}^{k_1,1}$ is well defined for each $k_1 \in [2, k+1]$. Moreover, $c_{\beta,\delta}^{k_1,1} < d_k$

for any $\delta \in (0, 2^{-1/4})$ and $k_1 \in [2, k+1]$. Define $\mathscr{N}_{b,\beta} := \{\mathbf{u} \in \mathscr{N}_b : J_\beta(\mathbf{u}) < d_k\}$ as in Sect. 3.2. Under the new definition (3.52), it is easy to see that Lemma 3.9 also holds here. Now as in Sect. 3.2, we define a map $V : \mathscr{N}_b^* \to H$ by $V(\mathbf{u}) := \mathbf{u} - K(\mathbf{u})$. Then Lemma 3.10 also holds here. Recall from (3.46) and (3.50) that $\int_\Omega (u_2^+)^3 (u_2 - w_2)\, \mathrm{d}x = 1 - 1 = 0$ for any $\mathbf{u} = (u_1, u_2) \in \mathscr{N}_b$. Then by similar arguments, we see that Lemma 3.11 also holds here.

Lemma 3.17 *There exists a unique global solution* $\eta = (\eta_1, \eta_2) : [0, \infty) \times \mathscr{N}_{b,\beta} \to H$ *for the initial value problem*

$$\frac{d}{\mathrm{d}t} \eta(t, \mathbf{u}) = -V(\eta(t, \mathbf{u})), \quad \eta(0, \mathbf{u}) = \mathbf{u} \in \mathscr{N}_{b,\beta}. \tag{3.53}$$

Furthermore, conclusions (i), (iii), *and* (iv) *of Lemma 3.12 also hold here, and* $\eta(t, \sigma_1(\mathbf{u})) = \sigma_1(\eta(t, \mathbf{u}))$ *for any* $t > 0$ *and* $u \in \mathscr{N}_{b,\beta}$.

Proof Recalling $V(\mathbf{u}) \in C^1(\mathscr{N}_b^*, H)$, we see that (3.53) has a unique solution η : $[0, T_{\max}) \times \mathscr{N}_{b,\beta} \to H$, where $T_{\max} > 0$ is the maximal time such that $\eta(t, \mathbf{u}) \in \mathscr{N}_b^*$ for all $t \in [0, T_{\max})$. Fix any $\mathbf{u} = (u_1, u_2) \in \mathscr{N}_{b,\beta}$, we deduce from (3.53) that

$$\frac{d}{\mathrm{d}t} \int_\Omega \left(\eta_2(t, \mathbf{u})^+\right)^4 \mathrm{d}x = 4 - 4 \int_\Omega \left(\eta_2(t, \mathbf{u})^+\right)^4 \mathrm{d}x, \quad \forall 0 < t < T_{\max}.$$

Since $\int_\Omega \left(\eta_2(0, \mathbf{u})^+\right)^4 \mathrm{d}x = \int_\Omega (u_2^+)^4 \mathrm{d}x = 1$, so $\int_\Omega \left(\eta_2(t, \mathbf{u})^+\right)^4 \mathrm{d}x \equiv 1$ for all $0 \le t < T_{\max}$. Recalling (3.51), the rest of the proof is similar to Lemma 3.12. $\square$

Now we can finish the proof Theorem 3.4.

Proof (Proof of Theorem 3.4 continued: II Existence of k semi-nodal solutions) First we fix any $k_1 \in [2, k+1]$. Then by similar arguments as Step 1 in the proof of Theorem 3.4, for small $\delta > 0$, there exists $\mathbf{u} = (u_1, u_2) \in \mathscr{N}_b$ such that

$$J_\beta(\mathbf{u}) = c_{\beta,\delta}^{k_1,1}, \quad V(\mathbf{u}) = 0 \quad \text{and} \quad \mathrm{dist}_4(\mathbf{u}, \mathscr{P}) \ge \delta.$$

Consequently, u_1 changes sign. Since $V(\mathbf{u}) = 0$, we have $\mathbf{u} = K(\mathbf{u})$. Combining this with (3.50), we see that $\mathbf{u}$ satisfies

$$\begin{cases} -\Delta u_1 + \lambda_1 u_1 = \alpha_1 \mu_1 t_1(\mathbf{u}) u_1^3 + \beta t_2(\mathbf{u}) u_2^2 u_1, \\ -\Delta u_2 + \lambda_2 u_2 = \alpha_2 \mu_2 t_2(\mathbf{u})(u_2^+)^3 + \beta t_1(\mathbf{u}) u_1^2 u_2. \end{cases} \tag{3.54}$$

Since $|u_1|_4 = 1$, $|u_2^+|_4 = 1$ and $t_i(\mathbf{u})$ satisfies (3.42), as before we have $\alpha_1 = \alpha_2 = 1$. Multiplying the second equation of (3.54) by u_2^- and integrating over Ω, we derive from (3.17) that $\|u_2^-\|_{\lambda_2}^2 = 0$, namely $u_2 \ge 0$. By the strong maximum principle, $u_2 > 0$ in Ω. Hence $(\tilde{u}_1, \tilde{u}_2) := (\sqrt{t_1(\mathbf{u})} u_1, \sqrt{t_2(\mathbf{u})} u_2)$ is a semi-nodal solution of the original problem (3.1) with $\tilde{u}_1$ sign-changing and $\tilde{u}_2$ positive. Moreover, (3.44) and (3.47) yield $E_\beta(\tilde{u}_1, \tilde{u}_2) = \widetilde{E}_\beta(\tilde{u}_1, \tilde{u}_2) = J_\beta(u_1, u_2) = c_{\beta,\delta}^{k_1,1} < d_k$. Finally, since $k_1 \in [2, k+1]$, by similar arguments as Step 2 of proving Theorem 3.4 with trivial modifications, we can prove that (3.1) has at least k semi-nodal solutions. $\square$

Remark 3.5 By a similar argument as in Sect. 3.2, we can prove that there exists $\beta_1'' > 0$ such that for any $\beta \in (0, \beta_1'')$, (3.1) has a semi-nodal solution which has the least energy among all semi-nodal solutions.

Chapter 4
A BEC System with Dimensions $N = 4$: Critical Case

Abstract As introduced in Chap. 1, we make a systematic study of the ground state solutions to system (1.2) in a smooth bounded domain $\Omega \subset \mathbb{R}^4$. An essential difference from the previous two chapters is that: In dimension 4, the cubic nonlinearities and coupling terms are all of critical growth, which make the study of system (1.2) much more challenging in view of mathematics. In this chapter, we prove the existence of ground state solutions for *almost full* ranges of parameter β. Furthermore, the uniqueness of ground state solutions can be obtained under some special situations. We also study the asymptotic behaviors of ground state solutions as $\beta \to -\infty$, where the so-called *phase separation* phenomena appears. Our proof is purely variational, where accurate energy estimates are established to prevent possible blow up of either one of the two components of a minimizing sequence. To the best of our knowledge, our study seems to be the first one for this BEC system in the critical case.

4.1 Main Results

As in Chaps. 2 and 3, we continue to study the BEC system (1.2). In Chaps. 2 and 3, we deal with the case $N = 2, 3$, where, in view of mathematics, the cubic nonlinearities and coupling terms are all of subcritical growth. In this chapter, we study system (1.2) in the critical case $N = 4$, namely

$$\begin{cases} -\Delta u + \lambda_1 u = \mu_1 u^3 + \beta u v^2, & x \in \Omega, \\ -\Delta v + \lambda_2 v = \mu_2 v^3 + \beta v u^2, & x \in \Omega, \\ u \geq 0, v \geq 0, & x \in \Omega, \\ u = v = 0, & x \in \partial\Omega, \end{cases} \tag{4.1}$$

where $\Omega \subset \mathbb{R}^4$ is a smooth bounded domain. Since $2^* := \frac{2N}{N-2} = 4$, it is known that the cubic nonlinearities and coupling terms are all of critical growth, and it follows that the Palais–Smale condition fails because of lacking the compactness property of the Sobolev embedding. This fact makes the study of system (1.2) much more tough comparing with that in the case $N = 2, 3$. Due to this reason, there is no any results

© Springer-Verlag Berlin Heidelberg 2015
Z. Chen, *Solutions of Nonlinear Schrödinger Systems*, Springer Theses,
DOI 10.1007/978-3-662-45478-7_4

about the critical exponent problem (4.1) in the literature. Therefore, our study here will be the *first* contribution to the BEC system (1.2) in the critical case.

Recall in Chap. 1 that $\lambda_1(\Omega)$ is the first eigenvalue of $-\Delta$ with the Dirichlet boundary condition, with its corresponding eigenfunction $\phi_1 > 0$. In this chapter, we always assume $-\lambda_1(\Omega) < \lambda_1, \lambda_2 < 0$ (the reason can be seen in Remark 4.2). It follows that system (4.1) has at least two semi-trivial solutions $(u_{\mu_1}, 0)$ and $(0, u_{\mu_2})$, where u_{μ_i} is a positive least energy solution of the well-known Brezis–Nirenberg problem

$$- \Delta u + \lambda_i u = \mu_i u^3, \quad u \geq 0 \text{ in } \Omega, \quad u|_{\partial\Omega} = 0. \tag{4.2}$$

See [19]. Therefore, system (4.1) can be also seen as a critically coupled perturbed Brezis–Nirenberg problem in dimension 4. Since the celebrated work [19] from Brezis and Nirenberg in 1983, the Brezis–Nirenberg problem has always been one of the focused topics in elliptic PDE in the past 30 years. For recent developments about this classical problem, we refer the reader to [30, 32, 43, 50, 79] and references therein.

Here, we are only interested in nontrivial solutions of system (4.1). Recall that $E_\beta, \mathcal{N}_\beta, A_\beta$ are defined in (1.4), (1.5), and (1.6), respectively. As before, when there is no confusions, we use $E, \mathcal{N}, A$ to denote $E_\beta, \mathcal{N}_\beta, A_\beta$ for convenience.

First, we consider the simple case $-\lambda_1(\Omega) < \lambda_1 = \lambda_2 = \lambda < 0$. We know from [19] that the Brezis–Nirenberg problem

$$- \Delta u + \lambda u = u^3, \quad u \geq 0 \text{ in } \Omega, \quad u|_{\partial\Omega} = 0 \tag{4.3}$$

admits a positive least energy solution ω, and its corresponding least energy is

$$B_1 := \frac{1}{4} \int_\Omega (|\vec{\nabla}\omega|^2 + \lambda\omega^2) \, dx = \frac{1}{4} \int_\Omega \omega^4 \, dx. \tag{4.4}$$

Moreover, it is easy to check that

$$\int_\Omega (|\nabla u|^2 + \lambda u^2) \, dx \geq 2\sqrt{B_1} \left(\int_\Omega u^4 \, dx \right)^{1/2}, \quad \forall \, u \in H_0^1(\Omega). \tag{4.5}$$

Our first result is following, which extend Theorem A in Chap. 2 to the critical case $N = 4$.

Theorem 4.1 *Let* $-\lambda_1(\Omega) < \lambda_1 = \lambda_2 = \lambda < 0$.

(1) *If* $0 < \beta < \min\{\mu_1, \mu_2\}$ *or* $\beta > \max\{\mu_1, \mu_2\}$, *then A is attained by* $(\sqrt{k}\omega, \sqrt{l}\omega)$, *where* $k, l > 0$ *satisfy*

$$\begin{cases} \mu_1 k + \beta l = 1, \\ \beta k + \mu_2 l = 1. \end{cases} \tag{4.6}$$

That is, $(\sqrt{k}\omega, \sqrt{l}\omega)$ is a positive ground state solution of (4.1).

(2) *If $\beta \in [\min\{\mu_1, \mu_2\}, \max\{\mu_1, \mu_2\}]$ and $\mu_1 \neq \mu_2$, then* (4.1) *has no nontrivial nonnegative solutions.*

We will see in Remark 4.2 that the assumption in Theorem 4.1 is optimal if Ω is a star-shaped domain. Now, we study the classification of ground state solutions.

Theorem 4.2 *Let $-\lambda_1(\Omega) < \lambda_1 = \lambda_2 = \lambda < 0$, and either $0 < \beta < \min\{\mu_1, \mu_2\}$ or $\beta > \max\{\mu_1, \mu_2\}$. Assume that (u, v) is a ground state solution of* (4.1), *then $(u, v) = (\sqrt{k}U, \sqrt{l}U)$, where (k, l) satisfies* (4.6) *and U is a least energy solution of the Brezis–Nirenberg problem* (4.3). *In particular, if Ω is an open ball in $\mathbb{R}^4$, then the ground state solution of* (4.1) *is unique.*

Now, we turn to consider the general case $-\lambda_1(\Omega) < \lambda_1, \lambda_2 < 0$, which is more delicate than the symmetric case $\lambda_1 = \lambda_2$. Without loss of generality, we may assume that $\lambda_1 \leq \lambda_2$. Our following result is more general, where we can also deal with the repulsive case $\beta < 0$.

Theorem 4.3 *Let $-\lambda_1(\Omega) < \lambda_1 \leq \lambda_2 < 0$.*

(1) *System* (4.1) *has a positive ground state solution (u, v) with $E(u, v) = A$ for any fixed $\beta < 0$.*

(2) *There exists $\beta_1 \in (0, \min\{\mu_1, \mu_2\})$, such that* (4.1) *has a positive ground state solution (u, v) with $E(u, v) = A$ for any fixed $\beta \in (0, \beta_1]$.*

(3) *Define*

$$\beta_2 := \max\left\{\frac{\lambda_1(\Omega) + \lambda_2}{\lambda_1(\Omega) + \lambda_1}\mu_1, \ \mu_2\right\}. \tag{4.7}$$

Then $\beta_2 \geq \max\{\mu_1, \mu_2\}$, and for any fixed $\beta > \beta_2$, system (4.1) *admits a positive ground state solution (u, v) with $E(u, v) = A$.*

(4) *If $\mu_2 \leq \beta \leq \mu_1$ and $\mu_2 < \mu_1$, then system* (4.1) *has no nontrivial positive solutions.*

Remark 4.1 Results in this chapter were published in a joint work with Zou [33]. Remark that, the expression (4.7) of β_2 here is *different* from that in [33], and the proof is also different. It seems difficult for us to say which expression is more optimal than the other one. In fact, we do not think that any of these two definitions of β_2 is optimal. As in Chap. 2, we can ask a fundamental question: What are the optimal ranges of parameter β for the existence of ground state solutions? This question is surely very important, but also very difficult, and remains open.

Remark 4.2 In fact, we can also give an accurate expression of the constant β_1 (see Lemma 4.2), but we do not give it here to avoid introducing heavy notations at this stage. Assume that $\beta > 0$. We multiply the equation for u in (4.1) by the first eigenfunction ϕ_1 and integrate over Ω, which yields

$$(\lambda_1 + \lambda_1(\Omega)) \int_\Omega u\phi_1 = \int_\Omega (\mu_1 u^3 \phi_1 + \beta uv^2 \phi_1) > 0.$$

Thus, we have to assume $\lambda_1, \lambda_2 > -\lambda_1(\Omega)$ since we want to obtain a nontrivial positive solution of (4.1). On the other hand, if Ω is star-shaped with respect to some x_0 and $\lambda_1, \lambda_2 \geq 0$, then by using the Pohozaev identity and $E'(u, v)(u, v) = 0$, it is easy to derive that

$$0 \leq \int_{\partial\Omega} (|\nabla u|^2 + |\nabla v|^2)((x - x_0) \cdot v) \, d\sigma = -2 \int_\Omega (\lambda_1 u^2 + \lambda_2 v^2) \, dx \leq 0,$$

where v denotes the exterior unit normal. This implies that $(u, v) \equiv (0, 0)$. This is one reason that we require the assumption $\lambda_1, \lambda_2 < 0$ in Theorems 4.1 and 4.3. This assumption is also needed in the proof of Lemma 4.2 in Sect. 4.4, because (4.1) is a critical exponent problem, and we always need to give appropriate upper bounds for the least energy to prove the strong convergence of a minimizing sequence. This idea was originally introduced by Brezis and Nirenberg [19].

Now, we study the asymptotic behaviors of the positive ground state solutions as $\beta \to -\infty$. It is expected that components of the limiting profile tend to repel each other and separate in different regions of the underlying domain Ω. That is, these solutions converge to a segregated limiting profile. This phenomenon, called *phase separation* in physics, has been well studied for L^∞-bounded positive solutions of system (4.1) in the case $N = 2, 3$ by [23, 73, 87, 88]. Denote $\{u > 0\} := \{x \in \Omega \mid u(x) > 0\}$. Then, we have the following result.

Theorem 4.4 *Let $-\lambda_1(\Omega) < \lambda_1 \leq \lambda_2 < 0$ as in Theorem 4.3. Let $\beta_n < 0$, $n \in \mathbb{N}$ satisfy $\beta_n \to -\infty$ as $n \to \infty$, and (u_n, v_n) be the positive ground state solutions of (4.1) with $\beta = \beta_n$. Then $\int_\Omega \beta_n u_n^2 v_n^2 \, dx \to 0$ as $n \to \infty$, and passing to a subsequence, one of the following conclusions holds.*

(1) $u_n \to u_\infty$ *strongly in $H_0^1(\Omega)$ and $v_n \rightharpoonup 0$ weakly in $H_0^1(\Omega)$ (so $v_n \to 0$ for almost every $x \in \Omega$), where u_∞ is a positive least energy solution of*

$$-\Delta u + \lambda_1 u = \mu_1 u^3, \quad u \in H_0^1(\Omega).$$

(2) $v_n \to v_\infty$ *strongly in $H_0^1(\Omega)$ and $u_n \rightharpoonup 0$ weakly in $H_0^1(\Omega)$ (so $u_n \to 0$ for almost every $x \in \Omega$), where v_∞ is a positive least energy solution of*

$$-\Delta v + \lambda_2 v = \mu_2 v^3, \quad v \in H_0^1(\Omega).$$

(3) $(u_n, v_n) \to (u_\infty, v_\infty)$ *strongly in $H_0^1(\Omega) \times H_0^1(\Omega)$ and $u_\infty \cdot v_\infty \equiv 0$, where $u_\infty \in C(\overline{\Omega})$ is a positive least energy solution of*

$$-\Delta u + \lambda_1 u = \mu_1 u^3, \ u \in H_0^1(\{u_\infty > 0\}),$$

and $v_\infty \in C(\overline{\Omega})$ is a positive least energy solution of

$$-\Delta v + \lambda_2 v = \mu_2 v^3, \ v \in H_0^1(\{v_\infty > 0\}).$$

Furthermore, both $\{v_\infty > 0\}$ and $\{u_\infty > 0\}$ are connected domains, and $\{v_\infty > 0\} = \Omega \backslash \overline{\{u_\infty > 0\}}$.

Remark 4.3 Under some further assumptions, we can exclude one of the three statements in Theorem 4.4. If $B_{\lambda_1, \mu_1} + \frac{1}{4} \mu_2^{-1} S^2 <$ (resp. $>$) $B_{\lambda_2, \mu_2} + \frac{1}{4} \mu_1^{-1} S^2$, then (2) (resp. (1)) in Theorem 4.4 does not hold, that is, either (1) (resp. (2)) or (3) holds. Here, S is defined in (4.9) and B_{λ_i, μ_i} denotes the least energy of (4.2) (see (4.23) in Sect. 4.4). The proof will be given in Sect. 4.5. For example, if we assume that $-\lambda_1(\Omega) < \lambda_1 < \lambda_2 < 0$ and $\mu_1 = \mu_2$ in Theorem 4.4, then $B_{\lambda_1, \mu_1} + \frac{1}{4} \mu_2^{-1} S^2 < B_{\lambda_2, \mu_2} + \frac{1}{4} \mu_1^{-1} S^2$, and so (2) in Theorem 4.4 does not hold.

As pointed out before, the nonlinearities and the coupling terms are all critical in (4.1). Therefore, the existence of nontrivial solutions of (4.1) depends heavily on the existence of the ground state solution of the following limit problem

$$\begin{cases} -\Delta u = \mu_1 u^3 + \beta u v^2, & x \in \mathbb{R}^4, \\ -\Delta v = \mu_2 v^3 + \beta v u^2, & x \in \mathbb{R}^4, \\ u, v \in D^{1,2}(\mathbb{R}^4), \end{cases} \tag{4.8}$$

where $D^{1,2}(\mathbb{R}^4) := \{u \in L^4(\mathbb{R}^4) : |\nabla u| \in L^2(\mathbb{R}^4)\}$ is the usual Sobolev space with norm $\|u\|_{D^{1,2}} := (\int_{\mathbb{R}^4} |\nabla u|^2 \, dx)^{1/2}$. Let S be the sharp constant of $D^{1,2}(\mathbb{R}^4) \hookrightarrow L^4(\mathbb{R}^4)$

$$\int_{\mathbb{R}^4} |\nabla u|^2 dx \geq S \left(\int_{\mathbb{R}^4} u^4 dx \right)^{\frac{1}{2}}. \tag{4.9}$$

For any $\varepsilon > 0$ and $y \in \mathbb{R}^4$, we define functions $U_{\varepsilon, y} \in D^{1,2}(\mathbb{R}^4)$ as

$$U_{\varepsilon, y}(x) := \frac{2\sqrt{2}\varepsilon}{\varepsilon^2 + |x - y|^2}. \tag{4.10}$$

Then $U_{\varepsilon, y}$ satisfies $-\Delta u = u^3$ in $\mathbb{R}^4$, and

$$\int_{\mathbb{R}^4} |\nabla U_{\varepsilon, y}|^2 \, dx = \int_{\mathbb{R}^4} |U_{\varepsilon, y}|^4 \, dx = S^2. \tag{4.11}$$

Furthermore, the set $\{U_{\varepsilon,y} \mid \varepsilon > 0, y \in \mathbb{R}^4\}$ contains all positive solutions of $-\Delta u = u^3$ in $\mathbb{R}^4$. See the Refs. [12, 82] for details.

Similarly as (4.1), the limit problem (4.8) also has two semi-trivial solutions $(\mu_1^{-1/2}U_{\varepsilon,y}, 0)$ and $(0, \mu_2^{-1/2}U_{\varepsilon,y})$. Here we are only interested in ground state solutions of (4.8). Denote $D := D^{1,2}(\mathbb{R}^4) \times D^{1,2}(\mathbb{R}^4)$ for simplicity and define a C^2 functional $I : D \to \mathbb{R}$ as

$$I(u, v) = \frac{1}{2}\int_{\mathbb{R}^4} |\nabla u|^2 + \frac{1}{2}\int_{\mathbb{R}^4} |\nabla v|^2 - \frac{1}{4}\int_{\mathbb{R}^4} (\mu_1 u^4 + 2\beta u^2 v^2 + \mu_2 v^4). \qquad (4.12)$$

Define the Nehari manifold of (4.8) as

$$\mathcal{M} = \left\{ (u, v) \in D \;\middle|\; u \neq 0, v \neq 0, \int_{\mathbb{R}^4} |\nabla u|^2 = \int_{\mathbb{R}^4} (\mu_1 u^4 + \beta u^2 v^2), \right.$$
$$\left. \int_{\mathbb{R}^4} |\nabla v|^2 = \int_{\mathbb{R}^4} (\mu_2 v^4 + \beta u^2 v^2) \right\}. \qquad (4.13)$$

Then all nontrivial solutions of (4.8) belong to $\mathcal{M}$. Similarly as $\mathcal{N}$, we can prove that $\mathcal{M} \neq \emptyset$. Define

$$B := \inf_{(u,v) \in \mathcal{M}} I(u, v) = \inf_{(u,v) \in \mathcal{M}} \frac{1}{4}\int_{\mathbb{R}^4} |\nabla u|^2 + |\nabla v|^2 \, dx. \qquad (4.14)$$

Similarly as Theorem 4.1, we can prove the following result for the limit problem (4.6), which plays a crucial role in the proof of Theorem 4.3.

Theorem 4.5 (1) *If $\beta < 0$, then B can not be attained.*

(2) *If $0 < \beta < \min\{\mu_1, \mu_2\}$ or $\beta > \max\{\mu_1, \mu_2\}$, then B is attained by the couple $(\sqrt{k}U_{\varepsilon,y}, \sqrt{l}U_{\varepsilon,y})$, where k, l can be seen in (4.6). That is, $(\sqrt{k}U_{\varepsilon,y}, \sqrt{l}U_{\varepsilon,y})$ are positive ground state solutions of (4.8).*

(3) *If $\beta \in [\min\{\mu_1, \mu_2\}, \max\{\mu_1, \mu_2\}]$ and $\mu_1 \neq \mu_2$, then (4.8) has no nontrivial positive solutions.*

In the rest of the sections of this chapter, we will prove the above five theorems. Theorems 4.1 and 4.5 are proved by following some ideas of [61, 80] in Sects. 4.2 and 4.3, respectively. In Sect. 4.4, we give the proof of Theorem 4.2 via a simple observation. Furthermore, the same argument also proves Theorem 2.2 of Chap. 2. In Sect. 4.5, we use Nehari manifold approach and Ekeland variational principle to prove (1)–(2) of Theorem 4.3, and use the classical mountain pass argument to prove (3) of Theorem 4.3. In Sect. 4.6, we use energy estimate methods to prove Theorem 4.4, where we need a powerful result from [73]. As before, we denote the norm of $L^p(\Omega)$ as $|u|_p = (\int_\Omega |u|^p \, dx)^{\frac{1}{p}}$, and the norm of $H_0^1(\Omega)$ as $\|u\| = |\nabla u|_2$.

4.2 The Simple Case $\lambda_1 = \lambda_2$

In this section, we give the proof of Theorem 4.1.

Proof (Proof of Theorem 4.1) Let $-\lambda_1(\Omega) < \lambda_1 = \lambda_2 = \lambda < 0$. Multiply the equation of u in (4.1) by v, the equation of v by u, and integrate over Ω, which implies

$$\int_{\Omega} uv[(\mu_1 - \beta)u^2 + (\beta - \mu_2)v^2] = 0.$$

Thus, (4.1) does not have nontrivial nonnegative solutions provided that

$$\beta \in [\min\{\mu_1, \mu_2\}, \max\{\mu_1, \mu_2\}], \quad \mu_1 \neq \mu_2,$$

namely (2) in Theorem 4.1 holds. It suffices to prove (1), and the idea of this proof comes from [80]. Since $0 < \beta < \min\{\mu_1, \mu_2\}$ or $\beta > \max\{\mu_1, \mu_2\}$, it follows that equation (4.6) has a solution (k, l) satisfying $k > 0$ and $l > 0$. Recalling (4.4), we see that $(\sqrt{k}\omega, \sqrt{l}\omega)$ is a nontrivial solution of (4.1) and

$$A \leq E(\sqrt{k}\omega, \sqrt{l}\omega) = (k + l)B_1. \tag{4.15}$$

Let $\{(u_n, v_n)\} \subset \mathcal{N}$ be a minimizing sequence for A, namely $E(u_n, v_n) \to A$. Define

$$c_n = \left(\int_{\Omega} u_n^4 \, dx\right)^{1/2}, \quad d_n = \left(\int_{\Omega} v_n^4 \, dx\right)^{1/2}.$$

Then by (4.5) we have

$$2\sqrt{B_1}c_n \leq \int_{\Omega} (|\nabla u_n|^2 + \lambda u_n^2) = \int_{\Omega} (\mu_1 u_n^4 + \beta u_n^2 v_n^2) \leq \mu_1 c_n^2 + \beta c_n d_n,$$

$$2\sqrt{B_1}d_n \leq \int_{\Omega} (|\nabla v_n|^2 + \lambda v_n^2) = \int_{\Omega} (\mu_2 v_n^4 + \beta u_n^2 v_n^2) \leq \mu_2 d_n^2 + \beta c_n d_n.$$

Since $E(u_n, v_n) = \frac{1}{4} \int_{\Omega} (\mu_1 u_n^4 + 2\beta u_n^2 v_n^2 + \mu_2 v_n^4)$, by (4.15) we have

$$2\sqrt{B_1}(c_n + d_n) \leq 4E(u_n, v_n) \leq 4(k + l)B_1 + o(1),$$

$$\mu_1 c_n + \beta d_n \geq 2\sqrt{B_1},$$

$$\beta c_n + \mu_2 d_n \geq 2\sqrt{B_1}.$$

By (4.6), the above three inequalities are equivalent to

$$\left(c_n - 2k\sqrt{B_1}\right) + \left(d_n - 2l\sqrt{B_1}\right) \leq o(1),$$
$$\mu_1\left(c_n - 2k\sqrt{B_1}\right) + \beta\left(d_n - 2l\sqrt{B_1}\right) \geq 0,$$
$$\beta\left(c_n - 2k\sqrt{B_1}\right) + \mu_2\left(d_n - 2l\sqrt{B_1}\right) \geq 0.$$

Consequently $c_n \to 2k\sqrt{B_1}$ and $d_n \to 2l\sqrt{B_1}$ as $n \to +\infty$, and then

$$4A = \lim_{n\to+\infty} 4E(u_n, v_n) \geq \lim_{n\to+\infty} 2\sqrt{B_1}(c_n + d_n) = 4(k + l)B_1.$$

Combining this with (4.15), we conclude

$$A = (k + l)B_1 = E(\sqrt{k}\omega, \sqrt{l}\omega), \tag{4.16}$$

and so $(\sqrt{k}\omega, \sqrt{l}\omega)$ is a positive ground state solution of (4.1).   □

4.3 The Limit Problem

In this section, we study the limit problem (4.6) and prove Theorem 4.5. For any $(u, v) \in \mathcal{M}$, (4.9) gives

$$\int_{\mathbb{R}^4} |\nabla u|^2 + |\nabla v|^2 = \int_{\mathbb{R}^4} (\mu_1 u^4 + \mu_2 v^4 + 2\beta u^2 v^2)$$

$$\leq C\int_{\mathbb{R}^4} (u^4 + v^4) \leq C\left(\int_{\mathbb{R}^4} (|\nabla u|^2 + |\nabla v|^2)\right)^2,$$

which implies

$$B = \inf_{(u,v)\in\mathcal{M}} \frac{1}{4}\int_{\mathbb{R}^4} |\nabla u|^2 + |\nabla v|^2 \, dx \geq C > 0. \tag{4.17}$$

Repeating the proof of Proposition A, we have the following result.

Lemma 4.1 (see [80]) *If A (resp. B) is attained by $(u, v) \in \mathcal{N}$ (resp. $(u, v) \in \mathcal{M}$), then (u, v) is a critical point of E (resp. I) provided $-\infty < \beta < \sqrt{\mu_1\mu_2}$.*

We are now in a position to prove Theorem 4.5.

Proof (Proof of Theorem 4.5) The proof of (3) in Theorem 4.5 is the same as Theorem 4.1. It suffices to prove (1) and (2).

(1) By (4.10) we know that $\omega_{\mu_i} := \mu_i^{-1/2} U_{1,0}$ satisfies equation $-\Delta u = \mu_i u^3$ in $\mathbb{R}^4$. Let $e_1 = (1, 0, 0, 0) \in \mathbb{R}^4$ and

$$(u_R(x), v_R(x)) = (\omega_{\mu_1}(x), \omega_{\mu_2}(x + Re_1)).$$

Then $v_R \rightharpoonup 0$ weakly in $D^{1,2}(\mathbb{R}^4) \cap L^4(\mathbb{R}^4)$. That is,

$$\lim_{R \to +\infty} \int_{\mathbb{R}^4} u_R^2 v_R^2 \leq \lim_{R \to +\infty} \left(\int_{\mathbb{R}^4} u_R^3 v_R \right)^{2/3} \left(\int_{\mathbb{R}^4} v_R^4 \right)^{1/3} = 0.$$

So for $R > 0$ sufficiently large, the equations

$$\begin{cases} \int_{\mathbb{R}^4} |\nabla u_R|^2 = \mu_1 \int_{\mathbb{R}^4} u_R^4 = t_{1,R}\mu_1 \int_{\mathbb{R}^4} u_R^4 + t_{2,R}\beta \int_{\mathbb{R}^4} u_R^2 v_R^2, \\ \int_{\mathbb{R}^4} |\nabla v_R|^2 = \mu_2 \int_{\mathbb{R}^4} v_R^4 = t_{2,R}\mu_2 \int_{\mathbb{R}^4} v_R^4 + t_{1,R}\beta \int_{\mathbb{R}^4} u_R^2 v_R^2, \end{cases}$$

have a solution $(t_{1,R}, t_{2,R})$ with

$$\lim_{R \to +\infty} (|t_{1,R} - 1| + |t_{2,R} - 1|) = 0.$$

Since $(\sqrt{t_{1,R}} u_R, \sqrt{t_{2,R}} v_R) \in \mathcal{M}$, we deduce from (4.11) that

$$B \leq I\left(\sqrt{t_{1,R}} u_R, \sqrt{t_{2,R}} v_R\right) = \frac{1}{4}\left(t_{1,R} \int_{\mathbb{R}^4} |\nabla u_R|^2 + t_{2,R} \int_{\mathbb{R}^4} |\nabla v_R|^2 \right)$$

$$= \frac{1}{4}\left(t_{1,R}\mu_1^{-1} + t_{2,R}\mu_2^{-1} \right) S^2.$$

Letting $R \to +\infty$, we conclude that $B \leq \frac{1}{4}(\mu_1^{-1} + \mu_2^{-1})S^2$.

On the other hand, for any $(u, v) \in \mathcal{M}$, it follows from $\beta < 0$ and (4.9) that

$$\int_{\mathbb{R}^4} |\nabla u|^2 \, dx \leq \mu_1 \int_{\mathbb{R}^4} u^4 \, dx \leq \mu_1 S^{-2} \left(\int_{\mathbb{R}^4} |\nabla u|^2 \right)^2,$$

so $\int_{\mathbb{R}^4} |\nabla u|^2 \, dx \geq \mu_1^{-1} S^2$. Similarly $\int_{\mathbb{R}^4} |\nabla v|^2 \, dx \geq \mu_2^{-1} S^2$. This, together with (4.14), gives $B \geq \frac{1}{4}(\mu_1^{-1} + \mu_2^{-1}) S^2$. Thus

$$B = \frac{1}{4}(\mu_1^{-1} + \mu_2^{-1}) S^2. \tag{4.18}$$

Assume by contradiction that B is attained by some $(u, v) \in \mathcal{M}$, then $(|u|, |v|) \in \mathcal{M}$ and $I(|u|, |v|) = B$. It follows from Lemma 4.1 that $(|u|, |v|)$ is a nontrivial solution of (4.8). By the maximum principle, $u > 0, v > 0$ and so $\int_{\mathbb{R}^4} u^2 v^2 > 0$. Consequently,

$$\int_{\mathbb{R}^4} |\nabla u|^2 \, dx < \mu_1 \int_{\mathbb{R}^4} u^4 \, dx \leq \mu_1 S^{-2} \left(\int_{\mathbb{R}^4} |\nabla u|^2 \right)^2.$$

By this, it is easy to prove that

$$B = I(u, v) = \frac{1}{4} \int_{\mathbb{R}^4} (|\nabla u|^2 + |\nabla v|^2) > \frac{1}{4}(\mu_1^{-1} + \mu_2^{-1}) S^2,$$

which yields a contradiction.

(2) The proof is similar to Theorem 4.1 in Sect. 4.2. As before, we know that $(\sqrt{k} U_{\varepsilon,y}, \sqrt{l} U_{\varepsilon,y})$ are solutions of (4.8) with

$$B \leq I\left(\sqrt{k} U_{\varepsilon,y}, \sqrt{l} U_{\varepsilon,y}\right) = \frac{1}{4}(k + l) S^2.$$

Let $\{(u_n, v_n)\} \subset \mathcal{M}$ be a minimizing sequence for B, namely $I(u_n, v_n) \to B$. As in the proof of Theorem 4.1, we define $c_n = \left(\int_{\mathbb{R}^4} u_n^4 \right)^{1/2}$ and $d_n = \left(\int_{\mathbb{R}^4} v_n^4 \right)^{1/2}$. Then

$$S c_n \leq \int_{\mathbb{R}^4} |\nabla u_n|^2 = \int_{\mathbb{R}^4} \mu_1 u_n^4 + \beta u_n^2 v_n^2 \leq \mu_1 c_n^2 + \beta c_n d_n,$$

$$S d_n \leq \int_{\mathbb{R}^4} |\nabla v_n|^2 = \int_{\mathbb{R}^4} \mu_2 v_n^4 + \beta u_n^2 v_n^2 \leq \mu_2 d_n^2 + \beta c_n d_n,$$

which is equivalent to

$$S(c_n + d_n) \leq 4I(u_n, v_n) \leq (k + l) S^2 + o(1),$$

$$\mu_1 c_n + \beta d_n \geq S, \quad \beta c_n + \mu_2 d_n \geq S.$$

Thus $c_n \to kS$ and $d_n \to lS$ as $n \to \infty$. Consequently

$$4B = \lim_{n \to +\infty} 4I(u_n, v_n) \geq \lim_{n \to +\infty} S(c_n + d_n) = (k + l)S^2,$$

which implies

$$B = \frac{1}{4}(k + l)S^2 = I\left(\sqrt{k}U_{\varepsilon,y}, \sqrt{l}U_{\varepsilon,y}\right). \tag{4.19}$$

Therefore, $(\sqrt{k}U_{\varepsilon,y}, \sqrt{l}U_{\varepsilon,y})$ are ground state solutions of (4.8). $\square$

4.4 Uniqueness of Ground State Solutions

In this section, we study the uniqueness of ground state solutions, and give the proof of Theorems 4.2 and 2.2. Let $-\lambda_1(\Omega) < \lambda_1 = \lambda_2 = \lambda < 0$. As pointed out in Sect. 2.1, Wei and Yao [89] proved some uniqueness results of positive solutions for system (2.1). In particular, [89, Theorem 4.2] proved the uniqueness of positive solutions for system (2.1) in the case $\beta > \max\{\mu_1, \mu_2\}$. We note that the argument of [89, Theorem 4.2] also works for (4.1). Thus, we refer the reader to [89, Theorem 4.2] for the proof of Theorem 4.2 in the case $\beta > \max\{\mu_1, \mu_2\}$. In the following, we only consider the case $0 < \beta < \min\{\mu_1, \mu_2\}$.

Proof (Proof of Theorem 4.2) Fix $\mu_1 > 0, \mu_2 > 0$ and $0 < \beta < \min\{\mu_1, \mu_2\}$. Let (u_0, v_0) be any a ground state solution of (4.1), then $u_0, v_0 > 0$ in Ω by the strong maximum principle. Recalling $(\sqrt{k}\omega, \sqrt{l}\omega)$ in Theorem 4.1, first we claim that

$$\int_\Omega u_0^4 \, dx = k^2 \int_\Omega \omega^4 \, dx. \tag{4.20}$$

Clearly, there exists $\delta > 0$ such that $0 < \beta < \min\{\mu, \mu_2\}$ for any $\mu \in (\mu_1 - \delta, \mu_1 + \delta)$. Then by Theorem 4.1, A is attained when μ_1 is replaced by μ in system (4.1). Recalling the definitions of $E, \mathcal{N}$ and A, they all depend on μ. Thus, we use notations $E_\mu, \mathcal{N}_\mu$ and $A(\mu)$ in this proof. By (4.6) and (4.16) we have

$$A(\mu) = \frac{\mu + \mu_2 - 2\beta}{\mu\mu_2 - \beta^2} B_1,$$

in particular, $A'(\mu_1) := \frac{d}{d\mu} A(\mu)|_{\mu_1}$ exists. Define

$$f(t, s, \mu) := t\mu \int_\Omega u_0^4 \, dx + s \int_\Omega \beta u_0^2 v_0^2 \, dx - \int_\Omega (|\nabla u_0|^2 + \lambda u_0^2) \, dx,$$

$$g(t, s, \mu) := s \int_\Omega \mu_2 v_0^4 \, dx + t \int_\Omega \beta u_0^2 v_0^2 \, dx - \int_\Omega (|\nabla v_0|^2 + \lambda v_0^2) \, dx,$$

then $f(1, 1, \mu_1) = g(1, 1, \mu_1) = 0$, and

$$\frac{\partial f}{\partial t}(1, 1, \mu_1) = \mu_1 \int_\Omega u_0^4 \, dx, \qquad \frac{\partial f}{\partial s}(1, 1, \mu_1) = \beta \int_\Omega u_0^2 v_0^2 \, dx,$$

$$\frac{\partial g}{\partial t}(1, 1, \mu_1) = \beta \int_\Omega u_0^2 v_0^2 \, dx, \qquad \frac{\partial g}{\partial s}(1, 1, \mu_1) = \mu_2 \int_\Omega v_0^4 \, dx.$$

Define a matrix

$$F := \begin{pmatrix} \dfrac{\partial f}{\partial t}(1, 1, \mu_1) & \dfrac{\partial f}{\partial s}(1, 1, \mu_1) \\[2mm] \dfrac{\partial g}{\partial t}(1, 1, \mu_1) & \dfrac{\partial g}{\partial s}(1, 1, \mu_1) \end{pmatrix}.$$

Clearly $\det(F) > 0$. Therefore, by the implicit function theorem, functions $t(\mu)$ and $s(\mu)$ are both well defined and class C^1 on $(\mu_1 - \delta_1, \mu_1 + \delta_1)$ for some $\delta_1 \leq \delta$. Moreover, $t(\mu_1) = s(\mu_1) = 1$, and so we may assume that $t(\mu), s(\mu) > 0$ for all $\mu \in (\mu_1 - \delta_1, \mu_1 + \delta_1)$ by choosing δ_1 smaller if necessary. By $f(t(\mu), s(\mu), \mu) \equiv g(t(\mu), s(\mu), \mu) \equiv 0$, it is easy to prove that

$$t'(\mu_1) = -\frac{\int_\Omega u_0^4 \int_\Omega \mu_2 v_0^4}{\det(F)}, \qquad s'(\mu_1) = \frac{\int_\Omega u_0^4 \int_\Omega \beta u_0^2 v_0^2}{\det(F)}.$$

By using the Taylor expansion, we obtain that $t(\mu) = 1 + t'(\mu_1)(\mu - \mu_1) + O((\mu - \mu_1)^2)$ and $s(\mu) = 1 + s'(\mu_1)(\mu - \mu_1) + O((\mu - \mu_1)^2)$. On the other hand, we note from $f(t(\mu), s(\mu), \mu) \equiv g(t(\mu), s(\mu), \mu) \equiv 0$ that $(\sqrt{t(\mu)}u_0, \sqrt{s(\mu)}v_0) \in \mathcal{N}_\mu$. Thus,

$$A(\mu) \leq E_\mu(\sqrt{t(\mu)}u_0, \sqrt{s(\mu)}v_0)$$

$$= \frac{1}{4}t(\mu) \int_\Omega (|\nabla u_0|^2 + \lambda u_0^2) \, dx + \frac{1}{4}s(\mu) \int_\Omega (|\nabla v_0|^2 + \lambda v_0^2) \, dx$$

$$= A(\mu_1) + \frac{1}{4}D(\mu - \mu_1) + O((\mu - \mu_1)^2),$$

where

$$D := t'(\mu_1) \int_\Omega (|\nabla u_0|^2 + \lambda u_0^2) + s'(\mu_1) \int_\Omega (|\nabla v_0|^2 + \lambda v_0^2)$$

$$= -\frac{\int_\Omega u_0^4 \int_\Omega \mu_2 v_0^4}{\det(F)} \int_\Omega (\mu_1 u_0^4 + \beta u_0^2 v_0^2) + \frac{\int_\Omega u_0^4 \int_\Omega \beta u_0^2 v_0^2}{\det(F)} \int_\Omega (\mu_2 v_0^4 + \beta u_0^2 v_0^2)$$

$$= -\int_\Omega u_0^4 \, dx.$$

Consequently, by letting $\mu \uparrow \mu_1$, we have $\frac{A(\mu)-A(\mu_1)}{\mu-\mu_1} \geq \frac{D}{4} + O((\mu - \mu_1))$, namely $A'(\mu_1) \geq \frac{D}{4}$. Similarly, $\frac{A(\mu)-A(\mu_1)}{\mu-\mu_1} \leq \frac{D}{4} + O((\mu-\mu_1))$ as $\mu \downarrow \mu_1$, namely $A'(\mu_1) \leq \frac{D}{4}$. Hence we conclude that

$$A'(\mu_1) = \frac{D}{4} = -\frac{1}{4} \int_\Omega u_0^4 \, dx.$$

On the other hand, Theorem 4.1 says that $(\sqrt{k}\omega, \sqrt{l}\omega)$ is also a ground state solution of (4.1), so we also have

$$A'(\mu_1) = -\frac{k^2}{4} \int_\Omega \omega^4 \, dx,$$

namely (4.20) holds.

By a similar argument, that is, by computing $A'(\mu_2)$ and $A'(\beta)$ respectively, we can prove that

$$\int_\Omega v_0^4 \, dx = l^2 \int_\Omega \omega^4 \, dx, \quad \int_\Omega u_0^2 v_0^2 \, dx = kl \int_\Omega \omega^4 \, dx.$$

Therefore,

$$\int_\Omega u_0^2 v_0^2 \, dx = \frac{l}{k} \int_\Omega u_0^4 \, dx = \frac{k}{l} \int_\Omega v_0^4 \, dx.$$

Define $(\tilde{u}, \tilde{v}) := (\frac{1}{\sqrt{k}} u_0, \frac{1}{\sqrt{l}} v_0)$. By (4.6) and $(u_0, v_0) \in \mathcal{N}$ we have

$$\int_\Omega |\nabla \tilde{u}|^2 + \lambda \tilde{u}^2 \, dx = \int_\Omega \tilde{u}^4 \, dx, \quad \int_\Omega |\nabla \tilde{v}|^2 + \lambda \tilde{v}^2 \, dx = \int_\Omega \tilde{v}^4 \, dx. \qquad (4.21)$$

Then by (4.5) we derive

$$\frac{1}{4} \int_\Omega |\nabla \tilde{u}|^2 + \lambda \tilde{u}^2 \, dx \geq B_1, \quad \frac{1}{4} \int_\Omega |\nabla \tilde{v}|^2 + \lambda \tilde{v}^2 \, dx \geq B_1, \qquad (4.22)$$

Thus,

$$A = (k+l)B_1 = \frac{1}{4} \int_\Omega (|\nabla u_0|^2 + \lambda u_0^2 + |\nabla v_0|^2 + \lambda v_0^2)$$

$$= \frac{1}{4}k \int_\Omega (|\nabla \tilde{u}|^2 + \lambda \tilde{u}^2) + \frac{1}{4}l \int_\Omega (|\nabla \tilde{v}|^2 + \lambda \tilde{v}^2)$$

$$\geq (k + l)B_1.$$

Consequently,

$$\frac{1}{4} \int_\Omega |\nabla \tilde{u}|^2 + \lambda \tilde{u}^2 \, dx = B_1, \quad \frac{1}{4} \int_\Omega |\nabla \tilde{v}|^2 + \lambda \tilde{v}^2 \, dx = B_1.$$

Combining this with (4.21), we conclude from [19] that $\tilde{u}$ and $\tilde{v}$ are both positive least energy solutions of (4.3). Since (u, v) satisfies (4.1), we have

$$-\Delta \tilde{u} + \lambda \tilde{u} = \mu_1 k \tilde{u}^3 + \beta l \tilde{u} \tilde{v}^2 = \tilde{u}^3,$$

namely $\tilde{u}\tilde{v}^2 = \tilde{u}^3$, so $\tilde{u} = \tilde{v}$. Denoting $U = \tilde{u}$, we conclude that $(u_0, v_0) = (\sqrt{k}U, \sqrt{l}U)$, where U is a positive least energy solution of (4.3).

Now, we assume that Ω is a ball in $\mathbb{R}^4$. Then, the least energy solution of the Brezis–Nirenberg problem (4.3) is unique (see [2] for instance). Therefore, the ground state solution of system (4.1) is unique. $\Box$

Proof (Proof of Theorem 2.2)   The proof is completely the same as that of Theorem 4.2. $\Box$

4.5 The General Case $\lambda_1 \neq \lambda_2$

In this section, we assume that $-\lambda_1(\Omega) < \lambda_1 \leq \lambda_2 < 0$ and prove Theorem 4.3. Repeating the proof of Lemma 2.2 in Chap. 2, we see that Theorem 4.3-(4) holds. In the following, we always assume $\beta \in \left(-\infty, \min\{\mu_1, \mu_2\}\right) \cup \left(\max\{\mu_1, \mu_2\}, +\infty\right)$. Recalling the definition of A in (1.6), since

$$\int_\Omega (|\nabla u|^2 + \lambda_i u^2) \geq \left(1 + \frac{\lambda_i}{\lambda_1(\Omega)}\right) \int_\Omega |\nabla u|^2, \quad i = 1, 2,$$

it follows easily that $A > 0$. As pointed out before, the Brezis–Nirenberg problem (4.2) has a positive least energy solution $u_{\mu_i} \in C^2(\Omega) \cap C(\overline{\Omega})$, and its corresponding least energy B_{μ_i} satisfies

$$\frac{1}{4}\left(\frac{\lambda_1(\Omega) + \lambda_i}{\lambda_1(\Omega)}\right)^2 \mu_i^{-1}S^2 \leq B_{\mu_i} := \frac{1}{2} \int_\Omega (|\nabla u_{\mu_i}|^2 + \lambda_i u_{\mu_i}^2) - \frac{1}{4} \int_\Omega \mu_i u_{\mu_i}^4$$

$$< \frac{1}{4}\mu_i^{-1}S^2, \quad i = 1, 2. \tag{4.23}$$

The following lemma gives accurate upper bounds of A, which plays the key role in seeking ground state solutions. In the proof we need the assumption $\lambda_1, \lambda_2 < 0$.

Lemma 4.2 (1) *Define*

$$\beta_1 := \min \left\{ \mu_1, \sqrt{\frac{\mu_1 \mu_2 B_{\mu_1}}{B_{\mu_2}}}, \sqrt{\frac{\mu_1 \mu_2 B_{\mu_2}}{B_{\mu_1}}}, \right.$$

$$\left. \mu_2 \frac{\lambda_1 + \lambda_1(\Omega)}{\lambda_2 + \lambda_1(\Omega)}, \frac{(\lambda_1 + \lambda_1(\Omega))(\lambda_2 + \lambda_1(\Omega))}{\lambda_1(\Omega)^2(\mu_1^{-1} + \mu_2^{-1})} \right\},$$

$$= \min \left\{ \sqrt{\frac{\mu_1 \mu_2 B_{\mu_1}}{B_{\mu_2}}}, \sqrt{\frac{\mu_1 \mu_2 B_{\mu_2}}{B_{\mu_1}}}, \frac{(\lambda_1 + \lambda_1(\Omega))(\lambda_2 + \lambda_1(\Omega))}{\lambda_1(\Omega)^2(\mu_1^{-1} + \mu_2^{-1})} \right\}.$$
$$\tag{4.24}$$

Then $\beta_1 < \min\{\mu_1, \mu_2\}$. Furthermore, for any fixed $\beta \in (0, \beta_1]$, there holds

$$A < \min \left\{ B_{\mu_1} + B_{\mu_2}, \ B \right\}. \tag{4.25}$$

(2) *If $\beta < 0$, then*

$$A < \min \left\{ B_{\mu_1} + \frac{1}{4}\mu_2^{-1}S^2, \ B_{\mu_2} + \frac{1}{4}\mu_1^{-1}S^2, \ B \right\}.$$

Remark 4.4 The repulsive case $\beta < 0$ is different from the attractive case $\beta > 0$. In fact, we may prove that

$$A > B_{\mu_1} + B_{\mu_2} \quad \text{if } \beta < 0.$$

The proof is as follows. Let $\beta < 0$. By Lemma 4.2 we can prove later that Theorem 4.3-(1) holds. Therefore, we may assume here that (4.1) has a positive ground state solution (u, v). Consequently, $\int_\Omega (|\nabla u|^2 + \lambda_1 u^2) < \int_\Omega \mu_1 u^4$. On the other hand, similarly as (4.5), we have

$$\int_\Omega (|\nabla u|^2 + \lambda_1 u^2) \, dx \geq 2\sqrt{B_{\mu_1}} \left(\int_\Omega \mu_1 u^4 \, dx \right)^{1/2}. \tag{4.26}$$

So $\int_\Omega (|\nabla u|^2 + \lambda_1 u^2) > 4B_{\mu_1}$. Similarly, $\int_\Omega (|\nabla v|^2 + \lambda_2 v^2) > 4B_{\mu_2}$. Thus,

$$A = E(u, v) = \frac{1}{4} \int_\Omega (|\nabla u|^2 + \lambda_1 u^2 + |\nabla v|^2 + \lambda_2 v^2) > B_{\mu_1} + B_{\mu_2}.$$

Now let us give the proof of Lemma 4.2.

Proof Define a matrix

$$F(u, v) := \begin{pmatrix} \mu_1 \int\limits_\Omega u^4 \, dx & \beta \int\limits_\Omega u^2 v^2 \, dx \\ \beta \int\limits_\Omega u^2 v^2 \, dx & \mu_2 \int\limits_\Omega v^4 \, dx \end{pmatrix}. \tag{4.27}$$

When $|F(u, v)| := \det F(u, v) > 0$, the inverse matrix of $F(u, v)$ is

$$F^{-1}(u, v) := \frac{1}{\det F(u, v)} \begin{pmatrix} \mu_2 \int\limits_\Omega v^4 \, dx & -\beta \int\limits_\Omega u^2 v^2 \, dx \\ -\beta \int\limits_\Omega u^2 v^2 \, dx & \mu_1 \int\limits_\Omega u^4 \, dx \end{pmatrix}. \tag{4.28}$$

(1) Assume that $\beta \in (0, \beta_1]$. Recalling $\lambda_1 \le \lambda_2$, it follows from the definition of β_1 that $\beta_1 < \min\{\mu_1, \mu_2\}$. Consequently $|F(u_{\mu_1}, u_{\mu_2})| > 0$. Since

$$\int\limits_\Omega (|\nabla u_{\mu_i}|^2 + \lambda_i u_{\mu_i}^2) = \int\limits_\Omega \mu_i u_{\mu_i}^4,$$

we conclude that $t_0 > 0$, $s_0 > 0$ satisfying $(\sqrt{t_0} u_{\mu_1}, \sqrt{s_0} u_{\mu_2}) \in \mathcal{N}$ is equivalent to

$$\begin{pmatrix} t_0 \\ s_0 \end{pmatrix} := F^{-1}(u_{\mu_1}, u_{\mu_2}) \begin{pmatrix} \int\limits_\Omega \mu_1 u_{\mu_1}^4 \\ \int\limits_\Omega \mu_2 u_{\mu_2}^4 \end{pmatrix}$$

$$= \frac{1}{|F(u_{\mu_1}, u_{\mu_2})|} \begin{pmatrix} \int\limits_\Omega \mu_2 u_{\mu_2}^4 (\int\limits_\Omega \mu_1 u_{\mu_1}^4 - \int\limits_\Omega \beta u_{\mu_1}^2 u_{\mu_2}^2) \\ \int\limits_\Omega \mu_1 u_{\mu_1}^4 (\int\limits_\Omega \mu_2 u_{\mu_2}^4 - \int\limits_\Omega \beta u_{\mu_1}^2 u_{\mu_2}^2) \end{pmatrix} > \begin{pmatrix} 0 \\ 0 \end{pmatrix}. \tag{4.29}$$

Here and in the following, $\begin{pmatrix} a \\ b \end{pmatrix} > \begin{pmatrix} 0 \\ 0 \end{pmatrix}$ means both $a > 0$ and $b > 0$.

On the other hand, we deduce from (4.23) to (4.24) that

$$\int\limits_\Omega \beta u_{\mu_1}^2 u_{\mu_2}^2 < \sqrt{\frac{\mu_1 \mu_2 B_{\mu_1}}{B_{\mu_2}}} \left(\int\limits_\Omega u_{\mu_1}^4 \right)^{1/2} \left(\int\limits_\Omega u_{\mu_2}^4 \right)^{1/2}$$

$$= 4 \sqrt{\frac{B_{\mu_1}}{B_{\mu_2}}} \sqrt{B_{\mu_1} B_{\mu_2}} = \int\limits_\Omega \mu_1 u_{\mu_1}^4.$$

Similarly, $\int_\Omega \beta u_{\mu_1}^2 u_{\mu_2}^2 < \int_\Omega \mu_2 u_{\mu_2}^4$. Thus (4.29) holds, that is, when (t_0, s_0) is defined by (4.29), we have $(\sqrt{t_0} u_{\mu_1}, \sqrt{s_0} u_{\mu_2}) \in \mathcal{N}$. Consequently

$$A \leq E(\sqrt{t_0}u_{\mu_1}, \sqrt{s_0}u_{\mu_2}) = \frac{t_0}{4} \int_\Omega (|\nabla u_{\mu_1}|^2 + \lambda_1 u_{\mu_1}^2) + \frac{s_0}{4} \int_\Omega (|\nabla u_{\mu_2}|^2 + \lambda_2 u_{\mu_2}^2)$$

$$= \frac{t_0}{4} \int_\Omega \mu_1 u_{\mu_1}^4 + \frac{s_0}{4} \int_\Omega \mu_2 u_{\mu_2}^4$$

$$< \frac{t_0}{4} \int_\Omega (\mu_1 u_{\mu_1}^4 + \beta u_{\mu_1}^2 u_{\mu_2}^2) + \frac{s_0}{4} \int_\Omega (\mu_2 u_{\mu_2}^4 + \beta u_{\mu_1}^2 u_{\mu_2}^2)$$

$$= \frac{1}{4} \int_\Omega (|\nabla u_{\mu_1}|^2 + \lambda_1 u_{\mu_1}^2) + \frac{1}{4} \int_\Omega (|\nabla u_{\mu_2}|^2 + \lambda_2 u_{\mu_2}^2)$$

$$= B_{\mu_1} + B_{\mu_2},$$

namely $A < B_{\mu_1} + B_{\mu_2}$.

It remains to prove $A < B$. Without loss of generality, we may assume that $0 \in \Omega$. Then there exists $\rho > 0$ such that $\{x : |x| \leq \rho\} \subset \Omega$. Let $\psi \in C_0^1(\Omega)$ be a nonnegative function with $\psi \equiv 1$ for $|x| \leq \rho$. Recalling $U_{\varepsilon,0}$ in (4.10) and (4.11), we define $U_\varepsilon := \psi U_{\varepsilon,0}$. Then by [19] or [90, Lemma 1.46], we have the following inequalities

$$\int_\Omega |\nabla U_\varepsilon|^2 = S^2 + O(\varepsilon^2), \quad \int_\Omega |U_\varepsilon|^4 = S^2 + O(\varepsilon^4),$$

$$\int_\Omega |U_\varepsilon|^2 \geq C\varepsilon^2 |\ln \varepsilon| + O(\varepsilon^2).$$

where C is a positive constant. Recalling k, l in (4.6), we define

$$(u_\varepsilon, v_\varepsilon) := (\sqrt{k} U_\varepsilon, \sqrt{l} U_\varepsilon). \tag{4.30}$$

Then we have

$$E\left(\sqrt{t}u_\varepsilon, \sqrt{s}v_\varepsilon\right) = \frac{1}{2}t \int_\Omega (|\nabla u_\varepsilon|^2 + \lambda_1 u_\varepsilon^2) + \frac{1}{2}s \int_\Omega (|\nabla v_\varepsilon|^2 + \lambda_2 v_\varepsilon^2)$$

$$- \frac{1}{4} \int_\Omega (t^2 \mu_1 u_\varepsilon^4 + 2ts\beta u_\varepsilon^2 v_\varepsilon^2 + s^2 \mu_2 v_\varepsilon^4)$$

$$\leq \frac{1}{2}(kt + ls)\left(S^2 - C\varepsilon^2 |\ln \varepsilon| + O(\varepsilon^2)\right)$$

$$- \frac{1}{4}\left(\mu_1 k^2 t^2 + 2\beta klts + \mu_2 l^2 s^2\right)\left(S^2 + O(\varepsilon^4)\right). \tag{4.31}$$

Denote

$$A_\varepsilon = S^2 - C\varepsilon^2 |\ln \varepsilon| + O(\varepsilon^2), \quad B_\varepsilon = S^2 + O(\varepsilon^4),$$

then $0 < A_\varepsilon < B_\varepsilon$ and $A_\varepsilon < S^2$ for $\varepsilon > 0$ small enough. Consider

$$f(t, s) := \frac{1}{2} A_\varepsilon (kt + ls) - \frac{1}{4} B_\varepsilon \left(\mu_1 k^2 t^2 + 2\beta klts + \mu_2 l^2 s^2 \right),$$

then it is easy to prove the existence of t_ε, $s_\varepsilon > 0$ such that

$$f(t_\varepsilon, s_\varepsilon) = \max_{t, s > 0} f(t, s).$$

Combining (4.6) with $\frac{\partial}{\partial t} f(t, s)|_{(t_\varepsilon, s_\varepsilon)} = \frac{\partial}{\partial s} f(t, s)|_{(t_\varepsilon, s_\varepsilon)} = 0$, we get

$$t_\varepsilon = s_\varepsilon = A_\varepsilon / B_\varepsilon.$$

Then it follows from (4.6), (4.19) and (4.31) that

$$\max_{t, s > 0} E(\sqrt{t} u_\varepsilon, \sqrt{s} v_\varepsilon) \leq \max_{t, s > 0} f(t, s)$$

$$= \frac{1}{2} (k + l) \frac{A_\varepsilon^2}{B_\varepsilon} - \frac{1}{4} (\mu_1 k^2 + 2\beta kl + \mu_2 l^2) \frac{A_\varepsilon^2}{B_\varepsilon}$$

$$= \frac{1}{4} (k + l) \frac{A_\varepsilon^2}{B_\varepsilon} < \frac{1}{4} (k + l) A_\varepsilon$$

$$< \frac{1}{4} (k + l) S^2 = B \quad \text{for } \varepsilon > 0 \text{ sufficiently small.} \quad (4.32)$$

(*We remark that* (4.32) *also holds for any* $\beta > \max\{\mu_1, \mu_2\}$ *since* (k, l) *is also well defined in this case*). Similarly as before, we have $|F(u_\varepsilon, v_\varepsilon)| > 0$. Furthermore, $\tilde{t}_\varepsilon > 0$, $\tilde{s}_\varepsilon > 0$ satisfy $(\sqrt{\tilde{t}_\varepsilon} u_\varepsilon, \sqrt{\tilde{s}_\varepsilon} v_\varepsilon) \in \mathcal{N}$ is equivalent to

$$\begin{pmatrix} \tilde{t}_\varepsilon \\ \tilde{s}_\varepsilon \end{pmatrix} = \frac{|U_\varepsilon|_4^4}{|F(u_\varepsilon, v_\varepsilon)|} \begin{pmatrix} kl^2 \left(\mu_2 \int_\Omega (|\nabla U_\varepsilon|^2 + \lambda_1 U_\varepsilon^2) - \beta \int_\Omega (|\nabla U_\varepsilon|^2 + \lambda_2 U_\varepsilon^2) \right) \\ k^2 l \left(\mu_1 \int_\Omega (|\nabla U_\varepsilon|^2 + \lambda_2 U_\varepsilon^2) - \beta \int_\Omega (|\nabla U_\varepsilon|^2 + \lambda_1 U_\varepsilon^2) \right) \end{pmatrix}$$

$$> \begin{pmatrix} 0 \\ 0 \end{pmatrix}. \quad (4.33)$$

Note that

$$\mu_2 \int_\Omega (|\nabla U_\varepsilon|^2 + \lambda_1 U_\varepsilon^2) - \beta \int_\Omega (|\nabla U_\varepsilon|^2 + \lambda_2 U_\varepsilon^2)$$

$$= (\mu_2 - \beta) \int_\Omega |\nabla U_\varepsilon|^2 + (\mu_2 \lambda_1 - \beta \lambda_2) \int_\Omega U_\varepsilon^2$$

$$> \lambda_1(\Omega)(\mu_2 - \beta) \int_\Omega U_\varepsilon^2 + (\mu_2\lambda_1 - \beta\lambda_2) \int_\Omega U_\varepsilon^2$$

$$= \left[\mu_2(\lambda_1(\Omega) + \lambda_1) - \beta(\lambda_1(\Omega) + \lambda_2)\right] \int_\Omega U_\varepsilon^2 > 0.$$

Similarly,

$$\mu_1 \int_\Omega (|\nabla U_\varepsilon|^2 + \lambda_2 U_\varepsilon^2) - \beta \int_\Omega (|\nabla U_\varepsilon|^2 + \lambda_1 U_\varepsilon^2)$$

$$> \left[\mu_1(\lambda_1(\Omega) + \lambda_2) - \beta(\lambda_1(\Omega) + \lambda_1)\right] \int_\Omega U_\varepsilon^2 > 0.$$

Thus (4.33) holds and then $(\sqrt{\tilde{t}_\varepsilon}u_\varepsilon, \sqrt{\tilde{s}_\varepsilon}v_\varepsilon) \in \mathcal{N}$ for $(\tilde{t}_\varepsilon, \tilde{s}_\varepsilon)$ defined in (4.33). Consequently

$$A \leq E\left(\sqrt{\tilde{t}_\varepsilon}u_\varepsilon, \sqrt{\tilde{s}_\varepsilon}v_\varepsilon\right) \leq \max_{t,s>0} E\left(\sqrt{t}u_\varepsilon, \sqrt{s}v_\varepsilon\right) < B.$$

This completes the proof of conclusion (1).

(2) Now, we consider the repulsive case $\beta < 0$. Let t_0 be the large root of equation

$$B_{\mu_1}t^2 - 4B_{\mu_1}t = \mu_2^{-1}s^2. \tag{4.34}$$

Then

$$B_{\mu_1}t^2 - 4B_{\mu_1}t > \mu_2^{-1}s^2, \quad \forall t > t_0. \tag{4.35}$$

Since $u_{\mu_1} \in C(\overline{\Omega})$ and $u_{\mu_1} \equiv 0$ on $\partial\Omega$, there exists

$$B(y_0, 2R) := \{x : |x - y_0| \leq 2R\} \subset \Omega,$$

such that

$$\delta := \max_{B(y_0,2R)} u_{\mu_1} \leq \min\left\{\frac{\mu_2}{2|\beta|}, \sqrt{\frac{|\lambda_2|}{2|\beta|t_0}}, \frac{\lambda_1 + \lambda_1(\Omega)}{2|\beta|}\right\}. \tag{4.36}$$

Take a nonnegative function $\psi \in C_0^1(B(y_0, 2R))$ such that $\psi \equiv 1$ for $|x - y_0| \leq R$. Define $v_\varepsilon = \psi U_{\varepsilon,y_0}$, where U_{ε,y_0} is seen in (4.10) and (4.11). Again by [19] or [90, Lemma 1.46], we have

$$\int_\Omega |\nabla v_\varepsilon|^2 = S^2 + O(\varepsilon^2), \quad \int_\Omega |v_\varepsilon|^4 = S^2 + O(\varepsilon^4), \tag{4.37}$$

$$\int_\Omega |v_\varepsilon|^2 \geq C\varepsilon^2 |\ln \varepsilon| + O(\varepsilon^2). \tag{4.38}$$

Noting that $\mathrm{supp}(v_\varepsilon) \subset B(y_0, 2R)$, we deduce from (4.36) that, for $t, s > 0$,

$$2|\beta|ts \int_\Omega u_{\mu_1}^2 v_\varepsilon^2 \leq 2|\beta|\delta ts \int_\Omega u_{\mu_1} v_\varepsilon^2 \leq |\beta|\delta t^2 \int_\Omega u_{\mu_1}^2 + |\beta|\delta s^2 \int_\Omega v_\varepsilon^4$$

$$\leq \frac{|\beta|\delta}{\lambda_1 + \lambda_1(\Omega)} t^2 \int_\Omega (|\nabla u_{\mu_1}|^2 + \lambda_1 u_{\mu_1}^2) + |\beta|\delta s^2 \int_\Omega v_\varepsilon^4$$

$$\leq \frac{1}{2} t^2 \int_\Omega \mu_1 u_{\mu_1}^4 + |\beta|\delta s^2 \int_\Omega v_\varepsilon^4, \tag{4.39}$$

so

$$E\left(\sqrt{t}u_{\mu_1}, \sqrt{s}v_\varepsilon\right) = \frac{1}{2}t \int_\Omega (|\nabla u_{\mu_1}|^2 + \lambda_1 u_{\mu_1}^2) + \frac{1}{2}s \int_\Omega (|\nabla v_\varepsilon|^2 + \lambda_2 v_\varepsilon^2)$$

$$- \frac{1}{4} \int_\Omega (t^2 \mu_1 u_{\mu_1}^4 + 2ts\beta u_{\mu_1}^2 v_\varepsilon^2 + s^2 \mu_2 v_\varepsilon^4)$$

$$\leq \frac{1}{2}t \int_\Omega (|\nabla u_{\mu_1}|^2 + \lambda_1 u_{\mu_1}^2) - \frac{1}{8}t^2 \int_\Omega \mu_1 u_{\mu_1}^4$$

$$+ \frac{1}{2}s \int_\Omega (|\nabla v_\varepsilon|^2 + \lambda_2 v_\varepsilon^2) - \frac{1}{4}s^2(\mu_2 - |\beta|\delta) \int_\Omega v_\varepsilon^4$$

$$= f(t) + g(s). \tag{4.40}$$

By (4.36), (4.37) and (4.38), we can obtain via a standard argument (see [19, 90] for instance) that

$$\max_{s>0} g(s) < \frac{1}{4}(\mu_2 - |\beta|\delta)^{-1}S^2 \leq \frac{1}{2\mu_2}S^2 \quad \text{for } \varepsilon > 0 \text{ small enough.} \tag{4.41}$$

On the other hand, (4.23) yields

$$f(t) = 2B_{\mu_1}t - \frac{1}{2}B_{\mu_1}t^2.$$

This, together with (4.35), gives

$$f(t) + g(s) < 0, \quad \forall t > t_0, \; s > 0,$$

Thus, we see from (4.40) that

$$\max_{t,s>0} E\left(\sqrt{t}u_{\mu_1}, \sqrt{s}v_\varepsilon\right) = \max_{0 < t \leq t_0, s > 0} E\left(\sqrt{t}u_{\mu_1}, \sqrt{s}v_\varepsilon\right). \tag{4.42}$$

For $0 < t \leq t_0, s > 0$, we derive from (4.36) that

$$|\beta|ts \int_\Omega u_{\mu_1}^2 v_\varepsilon^2 \leq |\beta|t_0 \delta^2 s \int_{B(y_0,2R)} v_\varepsilon^2 \leq -\frac{\lambda_2}{2}s \int_\Omega v_\varepsilon^2,$$

so

$$E\left(\sqrt{t}u_{\mu_1}, \sqrt{s}v_\varepsilon\right) = \frac{1}{2}t \int_\Omega (|\nabla u_{\mu_1}|^2 + \lambda_1 u_{\mu_1}^2) + \frac{1}{2}s \int_\Omega (|\nabla v_\varepsilon|^2 + \lambda_2 v_\varepsilon^2)$$

$$- \frac{1}{4} \int_\Omega (t^2 \mu_1 u_{\mu_1}^4 + 2ts\beta u_{\mu_1}^2 v_\varepsilon^2 + s^2 \mu_2 v_\varepsilon^4)$$

$$\leq \frac{1}{2}t \int_\Omega (|\nabla u_{\mu_1}|^2 + \lambda_1 u_{\mu_1}^2) - \frac{1}{4}t^2 \int_\Omega \mu_1 u_{\mu_1}^4$$

$$+ \frac{1}{2}s \int_\Omega (|\nabla v_\varepsilon|^2 + \frac{\lambda_2}{2}v_\varepsilon^2) - \frac{1}{4}s^2 \mu_2 \int_\Omega v_\varepsilon^4$$

$$= f_1(t) + g_1(s). \tag{4.43}$$

Note that $\max_{t>0} f_1(t) = f_1(1) = B_{\mu_1}$. Moreover, similarly as (4.41), we can obtain

$$\max_{s>0} g_1(s) < \frac{1}{4}\mu_2^{-1}S^2 \quad \text{for } \varepsilon \text{ small enough.}$$

Combining this with (4.42) and (4.43), we conclude

$$\max_{t,s>0} E\left(\sqrt{t}u_{\mu_1}, \sqrt{s}v_\varepsilon\right) = \max_{0 < t \leq t_0, s > 0} E\left(\sqrt{t}u_{\mu_1}, \sqrt{s}v_\varepsilon\right)$$

$$\leq \max_{t>0} f_1(t) + \max_{s>0} g_1(s)$$

$$< B_{\mu_1} + \frac{1}{4}\mu_2^{-1}S^2 \quad \text{for } \varepsilon \text{ small enough.} \tag{4.44}$$

On the other hand, similarly as (4.39), we have

$$\left(\int_\Omega \beta u_{\mu_1}^2 v_\varepsilon^2 \, dx \right)^2 \leq |\beta|^2 \delta^2 \left(\int_\Omega u_{\mu_1} v_\varepsilon^2 \, dx \right)^2$$

$$\leq |\beta|^2 \delta^2 \int_\Omega u_{\mu_1}^2 \, dx \int_\Omega v_\varepsilon^4 \, dx$$

$$\leq \frac{|\beta|^2 \delta^2}{(\lambda_1(\Omega) + \lambda_1)\mu_2} \int_\Omega \mu_1 u_{\mu_1}^4 \, dx \int_\Omega \mu_2 v_\varepsilon^4 \, dx$$

$$< \int_\Omega \mu_1 u_{\mu_1}^4 \, dx \int_\Omega \mu_2 v_\varepsilon^4 \, dx,$$

so (4.27) implies $|F(u_{\mu_1}, v_\varepsilon)| > 0$. Similarly as before, $t_\varepsilon > 0$, $s_\varepsilon > 0$ satisfying $(\sqrt{t_\varepsilon} u_{\mu_1}, \sqrt{s_\varepsilon} v_\varepsilon) \in \mathscr{N}$ is equivalent to

$$\binom{t_\varepsilon}{s_\varepsilon} := F^{-1}(u_{\mu_1}, v_\varepsilon) \left(\begin{array}{c} \int_\Omega |\nabla u_{\mu_1}|^2 + \lambda_1 u_{\mu_1}^2 \\ \int_\Omega |\nabla v_\varepsilon|^2 + \lambda_2 v_\varepsilon^2 \end{array} \right) > \binom{0}{0}. \tag{4.45}$$

Since $\beta < 0$, we see from (4.28) that every element of $F^{-1}(u_{\mu_1}, v_\varepsilon)$ is positive. Thus (4.45) holds, namely when $(t_\varepsilon, s_\varepsilon)$ is defined by (4.45), we have $(\sqrt{t_\varepsilon} u_{\mu_1}, \sqrt{s_\varepsilon} v_\varepsilon) \in \mathscr{N}$. Consequently

$$A \leq E(\sqrt{t_\varepsilon} u_{\mu_1}, \sqrt{s_\varepsilon} v_\varepsilon) \leq \max_{t,s>0} E(\sqrt{t} u_{\mu_1}, \sqrt{s} v_\varepsilon) < B_{\mu_1} + \frac{1}{4} \mu_2^{-1} S^2.$$

Similarly, we can prove $A < B_{\mu_2} + \frac{1}{4} \mu_1^{-1} S^2$. Finally, we deduce from (4.18) and (4.23) that

$$B > \max \left\{ B_{\mu_1} + \frac{1}{4} \mu_2^{-1} S^2, \ B_{\mu_2} + \frac{1}{4} \mu_1^{-1} S^2 \right\},$$

which completes the proof. $\qquad \square$

Clearly Lemma 4.2 yields the existence of $\varepsilon_\beta > 0$ such that, for any $\beta \leq \beta_1$,

$$A = A_\beta < \frac{1}{4}(\mu_1^{-1} + \mu_2^{-1} - \varepsilon_\beta) S^2.$$

Lemma 4.3 *Assume that $\beta \in (-\infty, \beta_1]$, where β_1 is defined in (4.24). Then there exist $C_2 > C_1 > 0$, such that for any $(u, v) \in \mathscr{N}$ with $E(u, v) < \frac{1}{4}(\mu_1^{-1} + \mu_2^{-1} - \varepsilon_\beta) S^2$, there holds*

$$C_1 \leq \int_\Omega u^4 \, dx, \int_\Omega v^4 \, dx \leq C_2.$$

Proof Fix any $(u, v) \in \mathcal{N}$ such that $E(u, v) < \frac{1}{4}(\mu_1^{-1} + \mu_2^{-1} - \varepsilon_\beta)S^2$. Since

$$\frac{\lambda_1(\Omega) + \lambda_1}{\lambda_1(\Omega)} S|u|_4^2 < \int_\Omega (|\nabla u|^2 + \lambda_1 u^2) = \int_\Omega (\mu_1 u^4 + \beta u^2 v^2) \leq \mu_1 |u|_4^4 + \beta_+ |u|_4^2 |v|_4^2,$$

$$\frac{\lambda_1(\Omega) + \lambda_2}{\lambda_1(\Omega)} S|v|_4^2 < \int_\Omega (|\nabla v|^2 + \lambda_2 v^2) = \int_\Omega (\mu_2 v^4 + \beta v^2 u^2) \leq \mu_2 |v|_4^4 + \beta_+ |u|_4^2 |v|_4^2,$$

where $\beta_+ = \max\{\beta, 0\}$. Hence there exists $C_2 > 0$ such that $\int_\Omega u^4, \int_\Omega v^4 \leq C_2$. Moreover, when $\beta \leq 0$, there exists $C > 0$ such that $\int_\Omega u^4, \int_\Omega v^4 \geq C$. It suffices to consider the case $\beta > 0$. Clearly

$$\mu_1 |u|_4^2 + \beta |v|_4^2 > \frac{\lambda_1(\Omega) + \lambda_1}{\lambda_1(\Omega)} S, \tag{4.46}$$

$$\beta |u|_4^2 + \mu_2 |v|_4^2 > \frac{\lambda_1(\Omega) + \lambda_2}{\lambda_1(\Omega)} S, \tag{4.47}$$

$$\frac{\lambda_1(\Omega) + \lambda_1}{\lambda_1(\Omega)} |u|_4^2 + \frac{\lambda_1(\Omega) + \lambda_2}{\lambda_1(\Omega)} |v|_4^2 < (\mu_1^{-1} + \mu_2^{-1} - \varepsilon_\beta)S. \tag{4.48}$$

Since $\lambda_1 \leq \lambda_2$ and $\beta \in (0, \beta_1]$, where the expression of β_1 is seen in (4.24), we deduce from (4.46) and (4.48) that

$$|u|_4^2 > \frac{S[(\lambda_1 + \lambda_1(\Omega))(\lambda_2 + \lambda_1(\Omega)) - \beta \lambda_1(\Omega)^2(\mu_1^{-1} + \mu_2^{-1} - \varepsilon_\beta)]}{\lambda_1(\Omega)[\mu_1(\lambda_2 + \lambda_1(\Omega)) - \beta(\lambda_1 + \lambda_1(\Omega))]} > 0,$$

Similarly, by (4.47) and (4.48), we have

$$|v|_4^2 > \frac{S[(\lambda_1 + \lambda_1(\Omega))(\lambda_2 + \lambda_1(\Omega)) - \beta \lambda_1(\Omega)^2(\mu_1^{-1} + \mu_2^{-1} - \varepsilon_\beta)]}{\lambda_1(\Omega)[\mu_2(\lambda_1 + \lambda_1(\Omega)) - \beta(\lambda_2 + \lambda_1(\Omega))]} > 0.$$

This completes the proof. $\square$

Now we can begin the proof of Theorem 4.3. First we consider the case $\beta \leq \beta_1$.

Proof (Proof of Theorem 4.3 (1)–(2)) Assume that $-\infty < \beta \leq \beta_1$ and $\beta \neq 0$. Note that E is coercive and bounded from below on $\mathcal{N}$. By the Ekeland variational principle (cf. [81]), there exists a minimizing sequence $\{(u_n, v_n)\} \subset \mathcal{N}$ of A satisfying

$$E(u_n, v_n) < \min\left\{A + \frac{1}{n}, \frac{1}{4}(\mu_1^{-1} + \mu_2^{-1} - \varepsilon_\beta)S^2\right\}, \tag{4.49}$$

$$E(u, v) \geq E(u_n, v_n) - \frac{1}{n}\|(u_n, v_n) - (u, v)\|, \quad \forall (u, v) \in \mathcal{N}. \tag{4.50}$$

Here $\|(u, v)\| := (\int_\Omega (|\nabla u|^2 + |\nabla v|^2)\, dx)^{1/2}$ is the norm of H. Clearly $\{(u_n, v_n)\}$ is uniformly bounded in H.

Define matrix

$$F_n := \begin{pmatrix} -\mu_1 \int\limits_\Omega u_n^4\, dx & -\beta \int\limits_\Omega u_n^2 v_n^2\, dx \\ -\beta \int\limits_\Omega u_n^2 v_n^2\, dx & -\mu_2 \int\limits_\Omega v_n^4\, dx \end{pmatrix}.$$

Then for $0 < \beta < \beta_1$, we derive from Lemma 4.3 that

$$\det(F_n) = \mu_1 \mu_2 \int\limits_\Omega u_n^4\, dx \int\limits_\Omega v_n^4\, dx - \beta^2 \left(\int\limits_\Omega u_n^2 v_n^2\, dx \right)^2$$

$$\geq (\mu_1 \mu_2 - \beta^2) \int\limits_\Omega u_n^4\, dx \int\limits_\Omega v_n^4\, dx \geq C > 0, \tag{4.51}$$

where C is independent of n. For $\beta < 0$, since $(u_n, v_n) \in \mathcal{N}$, we can also deduce from Lemma 4.3 that

$$\det(F_n) = \mu_1 \mu_2 \int\limits_\Omega u_n^4\, dx \int\limits_\Omega v_n^4\, dx - \beta^2 \left(\int\limits_\Omega u_n^2 v_n^2\, dx \right)^2$$

$$= \left(|\beta| \int\limits_\Omega u_n^2 v_n^2 + \int\limits_\Omega (|\nabla u_n|^2 + \lambda_1 |u_n|^2) \right)$$

$$\times \left(|\beta| \int\limits_\Omega u_n^2 v_n^2 + \int\limits_\Omega (|\nabla v_n|^2 + \lambda_2 |v_n|^2) \right) - \beta^2 \left(\int\limits_\Omega u_n^2 v_n^2\, dx \right)^2$$

$$\geq \int\limits_\Omega (|\nabla u_n|^2 + \lambda_1 |u_n|^2) \int\limits_\Omega (|\nabla v_n|^2 + \lambda_2 |v_n|^2)$$

$$\geq \frac{(\lambda_1(\Omega) + \lambda_1)(\lambda_1(\Omega) + \lambda_2)}{\lambda_1(\Omega)^2} S^2 |u_n|_4^2 |v_n|_4^2 \geq C > 0, \tag{4.52}$$

where C is independent of n. Thus $\det(F_n) \geq C > 0$ holds for all $-\infty < \beta < \beta_1$. Then by repeating the proof of Step 2 in Lemma 2.6, we conclude

$$\lim_{n \to +\infty} E'(u_n, v_n) = 0. \tag{4.53}$$

Since $\{(u_n, v_n)\}$ is bounded in H, we may assume that $(u_n, v_n) \rightharpoonup (u, v)$ weakly in H. Passing to a subsequence, we may assume that

$$u_n \rightharpoonup u, \quad v_n \rightharpoonup v, \quad \text{weakly in } L^4(\Omega),$$

$$u_n^2 \rightharpoonup u^2, \quad v_n^2 \rightharpoonup v^2, \quad \text{weakly in } L^2(\Omega),$$
$$u_n^3 \rightharpoonup u^3, \quad v_n^3 \rightharpoonup v^3, \quad \text{weakly in } L^{4/3}(\Omega),$$
$$u_n \to u, \quad v_n \to v, \quad \text{strongly in } L^2(\Omega).$$

Thus, by (4.53) we have $E'(u, v) = 0$. Set $\omega_n = u_n - u$ and $\sigma_n = v_n - v$. Then by Brezis–Lieb Lemma (cf. [90]), there holds

$$|u_n|_4^4 = |u|_4^4 + |\omega_n|_4^4 + o(1), \quad |v_n|_4^4 = |v|_4^4 + |\sigma_n|_4^4 + o(1). \tag{4.54}$$

On the other hand, since $\omega_n \rightharpoonup 0$ in $H_0^1(\Omega)$, passing to a subsequence, $\omega_n \to 0$ for almost every $x \in \Omega$. Therefore, we also have $|\omega_n| \rightharpoonup 0$ in $H_0^1(\Omega)$, and so

$$\omega_n^2 \rightharpoonup 0, \quad \sigma_n^2 \rightharpoonup 0, \quad \text{weakly in } L^2(\Omega),$$
$$|\omega_n|^3 \rightharpoonup 0, \quad |\sigma_n|^3 \rightharpoonup 0, \quad \text{weakly in } L^{4/3}(\Omega).$$

Consequently

$$\left| \int_\Omega u\omega_n v_n^2 \right| \leq \left(\int_\Omega |u||\omega_n|^3 \right)^{1/3} \left(\int_\Omega |u||v_n|^3 \right)^{2/3} = o(1),$$

$$\left| \int_\Omega u^2 \sigma_n(2v + \sigma_n) \right| \leq \left(\int_\Omega u^2 \sigma_n^2 \right)^{1/2} \left(\int_\Omega u^2(2v + \sigma_n)^2 \right)^{1/2} = o(1),$$

$$\left| \int_\Omega \omega_n^2 v(v + 2\sigma_n) \right| \leq \left(\int_\Omega |v||\omega_n|^3 \right)^{2/3} \left(\int_\Omega |v||(v + 2\sigma_n)|^3 \right)^{1/3} = o(1),$$

and so

$$\int_\Omega u_n^2 v_n^2 = \int_\Omega \omega_n^2 \sigma_n^2 + \int_\Omega u^2 v^2 + 2 \int_\Omega u\omega_n v_n^2$$

$$+ \int_\Omega u^2 \sigma_n(2v + \sigma_n) + \int_\Omega \omega_n^2 v(v + 2\sigma_n)$$

$$= \int_\Omega \omega_n^2 \sigma_n^2 + \int_\Omega u^2 v^2 + o(1). \tag{4.55}$$

Note that $(u_n, v_n) \in \mathcal{N}$ and $E'(u, v) = 0$. Combining these with (4.54) and (4.55), we obtain

$$\int_\Omega |\nabla \omega_n|^2 - \int_\Omega (\mu_1 \omega_n^4 + \beta \omega_n^2 \sigma_n^2) = o(1). \tag{4.56}$$

$$\int_\Omega |\nabla \sigma_n|^2 - \int_\Omega (\mu_2 \sigma_n^4 + \beta \omega_n^2 \sigma_n^2) = o(1), \tag{4.57}$$

$$E(u_n, v_n) = E(u, v) + I(\omega_n, \sigma_n) + o(1). \tag{4.58}$$

Up to a subsequence, we may assume that

$$\lim_{n \to +\infty} \int_\Omega |\nabla \omega_n|^2 = b_1, \quad \lim_{n \to +\infty} \int_\Omega |\nabla \sigma_n|^2 = b_2.$$

Then by (4.56) and (4.57), we have $I(\omega_n, \sigma_n) = \frac{1}{4}(b_1 + b_2) + o(1)$. Letting $n \to \infty$ in (4.58), we obtain

$$0 \le E(u, v) \le E(u, v) + \frac{1}{4}(b_1 + b_2) = \lim_{n \to +\infty} E(u_n, v_n) = A. \tag{4.59}$$

Case 1. $u \equiv 0, v \equiv 0$.

By Lemma 4.3 and (4.54), we have $b_1 > 0$ and $b_2 > 0$, so we may assume that both $\omega_n \not\equiv 0$ and $\sigma_n \not\equiv 0$ for n large. Then by (4.56) and (4.57), it is easy to prove the existence of $t_n, s_n > 0$ such that $(\sqrt{t_n}\omega_n, \sqrt{s_n}\sigma_n) \in \mathcal{M}$ and

$$\lim_{n \to +\infty} (|t_n - 1| + |s_n - 1|) = 0.$$

Therefore,

$$\frac{1}{4}(b_1 + b_2) = \lim_{n \to +\infty} I(\omega_n, \sigma_n) = \lim_{n \to +\infty} I(\sqrt{t_n}\omega_n, \sqrt{s_n}\sigma_n) \ge B.$$

Combining this with (4.59) we get that $A \ge B$, a contradiction with Lemma 4.2. Therefore, Case 1 is impossible.

Case 2. Either $u \not\equiv 0, v \equiv 0$ or $u \equiv 0, v \not\equiv 0$.

Without loss of generality, we may assume that $u \not\equiv 0, v \equiv 0$. Then $b_2 > 0$. By Case 1 we may assume that $b_1 = 0$. Then $\lim_{n \to +\infty} \int_\Omega \omega_n^2 \sigma_n^2 = 0$, and so

$$\int_\Omega |\nabla \sigma_n|^2 = \int_\Omega \mu_2 \sigma_n^4 + o(1) \le \mu_2 S^{-2} \left(\int_\Omega |\nabla \sigma_n|^2 \right)^2 + o(1).$$

This implies that $b_2 \ge \mu_2^{-1} S^2$. Noting that u is a nontrivial solution of $-\Delta u + \lambda_1 u = \mu_1 u^3$, we have from (4.23) that $E(u, 0) \ge B_{\mu_1}$. By (4.59) we get that

$$A \ge B_{\mu_1} + \frac{1}{4} b_2 \ge B_{\mu_1} + \frac{1}{4}\mu_2^{-1} S^2 > B_{\mu_1} + B_{\mu_2},$$

a contradiction with Lemma 4.2. Therefore, Case 2 is impossible.

Since Cases 1 and 2 are both impossible, we have that $u \not\equiv 0$ and $v \not\equiv 0$, namely $(u, v) \in \mathcal{N}$. By (4.59) we have $E(u, v) = A$. Consequently $(|u|, |v|) \in \mathcal{N}$ and $E(|u|, |v|) = A$. By Lemma 4.1, we conclude that $(|u|, |v|)$ is a solution of (4.1). Finally, the maximum principle gives that $|u|, |v| > 0$ in Ω. Therefore, $(|u|, |v|)$ is a positive ground state solution of (4.1). This completes the proof. $\qquad\square$

Now, we turn to prove Theorem 4.3-(3). In the following, we assume $\beta > \max\{\mu_1, \mu_2\}$. Define the mountain pass type minimax value

$$\mathscr{A} := \inf_{h \in \Gamma} \max_{t \in [0,1]} E(h(t)), \tag{4.60}$$

where $\Gamma = \{h \in C([0, 1], H) \mid h(0) = (0, 0), E(h(1)) < 0\}$. By (1.4) we know that for any $(u, v) \in H \setminus \{(0, 0)\}$,

$$\max_{t>0} E\left(\sqrt{t}u, \sqrt{t}v\right) = E\left(\sqrt{t_{u,v}}u, \sqrt{t_{u,v}}v\right)$$
$$= \frac{1}{4} t_{u,v} \int_\Omega (|\nabla u|^2 + \lambda_1 u^2 + |\nabla v|^2 + \lambda_2 v^2), \tag{4.61}$$

where $t_{u,v} > 0$ satisfies

$$t_{u,v} = \frac{\int_\Omega (|\nabla u|^2 + \lambda_1 u^2 + |\nabla v|^2 + \lambda_2 v^2)}{\int_\Omega (\mu_1 u^4 + 2\beta u^2 v^2 + \mu_2 v^4)}. \tag{4.62}$$

Note that $(\sqrt{t_{u,v}}u, \sqrt{t_{u,v}}v) \in \mathcal{N}'$, where

$$\mathcal{N}' := \left\{ (u, v) \in H \setminus \{(0, 0)\} \;\middle|\; G(u, v) := \int_\Omega (|\nabla u|^2 + \lambda_1 u^2 + |\nabla v|^2 + \lambda_2 v^2) \right.$$
$$\left. - \int_\Omega (\mu_1 u^4 + 2\beta u^2 v^2 + \mu_2 v^4) = 0 \right\}, \tag{4.63}$$

It is easy to check that

$$\mathscr{A} = \inf_{H \ni (u,v) \neq (0,0)} \max_{t>0} E\left(\sqrt{t}u, \sqrt{t}v\right) = \inf_{(u,v) \in \mathcal{N}'} E(u, v). \tag{4.64}$$

Since $\mathcal{N} \subset \mathcal{N}'$, we have $\mathscr{A} \leq A$. Similarly as (4.17), we also have $\mathscr{A} > 0$. Moreover, (4.30) and (4.32) give

$$0 < \mathscr{A} \leq \max_{t>0} E(\sqrt{t}u_\varepsilon, \sqrt{t}v_\varepsilon) \leq \max_{t,s>0} E(\sqrt{t}u_\varepsilon, \sqrt{s}v_\varepsilon) < B. \tag{4.65}$$

Lemma 4.4 *Recall the definition of β_2 in (4.7). Then $\beta_2 \geq \max\{\mu_1, \mu_2\}$, and for any $\beta > \beta_2$, there holds*

$$\mathscr{A} < \min\{B_{\mu_1}, B_{\mu_2}\}. \tag{4.66}$$

Proof Recalling (4.61)–(4.62), we define $t(s) := t_{u_{\mu_1}, \sqrt{s}u_{\mu_1}}$ for $s \geq 0$, namely

$$t(s) = \frac{\int_\Omega (|\nabla u_{\mu_1}|^2 + \lambda_1 u_{\mu_1}^2 + s|\nabla u_{\mu_1}|^2 + s\lambda_2 u_{\mu_1}^2)}{\int_\Omega (\mu_1 u_{\mu_1}^4 + 2s\beta u_{\mu_1}^4 + s^2 \mu_2 u_{\mu_1}^4)}.$$

Remark that $t(0) = 1$. A direct computation gives

$$\lim_{s \to 0} t'(s) = \frac{\int_\Omega (|\nabla u_{\mu_1}|^2 + \lambda_2 u_{\mu_1}^2 - 2\beta u_{\mu_1}^4)}{\int_\Omega \mu_1 u_{\mu_1}^4},$$

namely

$$t(s) = 1 + \frac{\int_\Omega (|\nabla u_{\mu_1}|^2 + \lambda_2 u_{\mu_1}^2 - 2\beta u_{\mu_1}^4)}{\int_\Omega \mu_1 u_{\mu_1}^4} s(1 + o(1)), \quad \text{for } s \to 0.$$

This implies

$$t^2(s) = 1 + \frac{2\int_\Omega (|\nabla u_{\mu_1}|^2 + \lambda_2 u_{\mu_1}^2 - 2\beta u_{\mu_1}^4)}{\int_\Omega \mu_1 u_{\mu_1}^4} s(1 + o(1)), \quad \text{for } s \to 0. \tag{4.67}$$

Remark that

$$\beta > \beta_2 := \max\left\{\frac{\lambda_1(\Omega) + \lambda_2}{\lambda_1(\Omega) + \lambda_1}\mu_1, \ \mu_2\right\} \geq \max\{\mu_1, \mu_2\},$$

we have

$$\int_\Omega (|\nabla u_{\mu_1}|^2 + \lambda_2 u_{\mu_1}^2 - \beta u_{\mu_1}^4)$$

$$= (\lambda_2 - \lambda_1)\int_\Omega u_{\mu_1}^2 + (\mu_1 - \beta)\int_\Omega u_{\mu_1}^4$$

$$\leq \frac{\lambda_2 - \lambda_1}{\lambda_1(\Omega) + \lambda_1}\int_\Omega (|\nabla u_{\mu_1}|^2 + \lambda_1 u_{\mu_1}^2) + (\mu_1 - \beta)\int_\Omega u_{\mu_1}^4$$

$$= \left(\frac{\lambda_1(\Omega) + \lambda_2}{\lambda_1(\Omega) + \lambda_1}\mu_1 - \beta\right)\int_\Omega u_{\mu_1}^4 < 0.$$

Thus, we deduce from (4.64) and (4.67) that

$$
\begin{aligned}
\mathscr{A} &\leq E\left(\sqrt{t(s)}u_{\mu_1}, \sqrt{t(s)}\sqrt{s}u_{\mu_1}\right) \\
&= \frac{1}{4}t^2(s)\int_\Omega (\mu_1 u_{\mu_1}^4 + 2s\beta u_{\mu_1}^4 + s^2\mu_2 u_{\mu_1}^4) \\
&= \frac{1}{4}\int_\Omega \mu_1 u_{\mu_1}^4 + \frac{s}{2}\int_\Omega (|\nabla u_{\mu_1}|^2 + \lambda_2 u_{\mu_1}^2 - \beta u_{\mu_1}^4) + O(s^2) \\
&< \frac{1}{4}\int_\Omega \mu_1 u_{\mu_1}^4 = B_{\mu_1}, \quad \text{for } s > 0 \text{ small enough.}
\end{aligned}
$$

Similarly, we can prove $\mathscr{A} < B_{\mu_2}$. $\qquad\square$

As pointed out in Sect. 4.1, we can give a *different* expression of β_2 to guarantee the validity of Lemma 4.4. For this, we refer the reader to our paper [33]. Now we can finish the proof of Theorem 4.3.

Proof (Proof of Theorem 4.3 (3)) Assume that $\beta > \beta_2$. Since the functional E has a mountain pass structure, by the classical mountain pass theorem (cf. [10, 90]), there exists $\{(u_n, v_n)\} \subset H$ such that

$$
\lim_{n\to+\infty} E(u_n, v_n) = \mathscr{A}, \quad \lim_{n\to+\infty} E'(u_n, v_n) = 0.
$$

Clearly $\{(u_n, v_n)\}$ is bounded in H, so we may assume that $(u_n, v_n) \rightharpoonup (u, v)$ weakly in H. Setting $\omega_n = u_n - u$ and $\sigma_n = v_n - v$ and using the same symbols as in the proof of (1)–(2) of Theorem 4.3, we see that $E'(u, v) = 0$ and (4.56)–(4.58) also hold. Moreover,

$$
0 \leq E(u, v) \leq E(u, v) + \frac{1}{4}(b_1 + b_2) = \lim_{n\to+\infty} E(u_n, v_n) = \mathscr{A} < B. \tag{4.68}
$$

Case 1. $u \equiv 0, v \equiv 0$.

By (4.68) we have $b_1 + b_2 > 0$. Similarly as the proof of Theorem 4.5-(2), we define $c_n = \left(\int_{\mathbb{R}^4} \omega_n^4\right)^{1/2}$ and $d_n = \left(\int_{\mathbb{R}^4} \sigma_n^4\right)^{1/2}$. If, up to a subsequence, $c_n \to 0$, then we can repeat the argument of Case 2 in the proof of Theorem 4.3 (1)–(2) to obtain that $b_2 \geq \mu_2^{-1}S^2$. Consequently

$$
\mathscr{A} \geq \frac{1}{4}b_2 \geq \frac{1}{4}\mu_2^{-1}S^2 > B_{\mu_2},
$$

a contradiction with Lemma 4.4. Similarly, $d_n \to 0$ also yields a contradiction. Thus, both c_n and d_n are uniformly bounded away from 0. By (4.56)–(4.57) we have

$$Sc_n \leq \int_{\mathbb{R}^4} |\nabla \omega_n|^2 = \int_{\mathbb{R}^4} \mu_1 \omega_n^4 + \beta \omega_n^2 \sigma_n^2 + o(1) \leq \mu_1 c_n^2 + \beta c_n d_n + o(1),$$

$$Sd_n \leq \int_{\mathbb{R}^4} |\nabla \sigma_n|^2 = \int_{\mathbb{R}^4} \mu_2 \sigma_n^4 + \beta \omega_n^2 \sigma_n^2 + o(1) \leq \mu_2 d_n^2 + \beta c_n d_n + o(1).$$

Combining these with (4.68) and (4.19), we obtain

$$S(c_n + d_n) \leq 4(b_1 + b_2) + o(1) \leq (k + l)S^2 + o(1),$$
$$\mu_1 c_n + \beta d_n + o(1) \geq S, \quad \beta c_n + \mu_2 d_n + o(1) \geq S.$$

Thus, $c_n \to kS$ and $d_n \to lS$ as $n \to \infty$, and then

$$4\mathscr{A} = 4(b_1 + b_2) \geq \lim_{n \to +\infty} S(c_n + d_n) = (k + l)S^2 = 4B,$$

which yields a contradiction with (4.65). So Case 1 is impossible.

Case 2. Either $u \not\equiv 0, v \equiv 0$ or $u \equiv 0, v \not\equiv 0$.

Without loss of generality, we may assume $u \not\equiv 0$ and $v \equiv 0$. Then u is a nontrivial solution of $-\Delta u + \lambda_1 u = \mu_1 u^3$, which implies $\mathscr{A} \geq E(u, 0) \geq B_{\mu_1}$, a contradiction with Lemma 4.4. So Case 2 is also impossible.

Since Cases 1 and 2 are both impossible, we have that both $u \not\equiv 0$ and $v \not\equiv 0$. Since $E'(u, v) = 0$, we have $(u, v) \in \mathscr{N}$. By (4.65) and (4.68) we have $E(u, v) = \mathscr{A} = A$. This means $(|u|, |v|) \in \mathscr{N} \subset \mathscr{N}'$ and $E(|u|, |v|) = \mathscr{A} = A$. By (4.63) and (4.64), there exists a Lagrange multiplier $\gamma \in \mathbb{R}$ such that

$$E'(|u|, |v|) - \gamma G'(|u|, |v|) = 0.$$

Since $E'(|u|, |v|)(|u|, |v|) = G(|u|, |v|) = 0$ and

$$G'(|u|, |v|)(|u|, |v|) = -2 \int_{\Omega} (\mu_1 u^4 + 2\beta u^2 v^2 + \mu_2 v^4) \neq 0,$$

we get that $\gamma = 0$ and so $E'(|u|, |v|) = 0$. This means that $(|u|, |v|)$ is a ground state solution of (4.1). By the maximum principle, we see that $|u|, |v| > 0$ in Ω. Therefore, $(|u|, |v|)$ is a positive ground state solution of (4.1). This completes the proof. $\qquad \square$

4.6 Phase Separation

In this section, we study the asymptotic behaviors of ground state solutions as $\beta \to -\infty$ and give the proof of Theorem 4.4. Recalling the definition of $E, \mathscr{N}$ and A in Chap. 1, we use notations $E_\beta, \mathscr{N}_\beta, A_\beta$ in the following proof.

Proof (Proof of Theorem 4.4) Let $\beta_n < 0$, $n \in \mathbb{N}$ satisfy $\beta_n \to -\infty$ as $n \to \infty$, and (u_n, v_n) be the positive ground state solutions of (4.1) with $\beta = \beta_n$. By Lemma 4.2, $E_{\beta_n}(u_n, v_n) \leq B = \frac{1}{4}(\mu_1^{-1} + \mu_2^{-1})S^2$ and so (u_n, v_n) is uniformly bounded in H by (1.6). Passing to a subsequence, we may assume that

$$u_n \rightharpoonup u_\infty, \quad v_n \rightharpoonup v_\infty \quad \text{weakly in } H_0^1(\Omega),$$

$$u_n \to u_\infty, \quad v_n \to v_\infty \quad \text{strongly in } L^2(\Omega),$$

$$u_n \to u_\infty, \quad v_n \to v_\infty \quad \text{a.e. } x \in \Omega.$$

Consequently $u_\infty, v_\infty \geq 0$ for almost every $x \in \Omega$.

Case 1. $u_\infty \equiv 0$, $v_\infty \equiv 0$.

Then $\int_\Omega u_n^2 \to 0$. Since $(u_n, v_n) \in \mathcal{N}_{\beta_n}$ and $\beta_n < 0$, we have

$$\int_\Omega |\nabla u_n|^2 + \lambda_1 \int_\Omega u_n^2 \leq \int_\Omega \mu_1 u_n^4 \leq \mu_1 S^{-2} \left(\int_\Omega |\nabla u_n|^2 \right)^2,$$

so

$$b_1 := \lim_{n \to \infty} \int_\Omega |\nabla u_n|^2 \geq \mu_1^{-1} S^2.$$

Similarly,

$$b_2 := \lim_{n \to \infty} \int_\Omega |\nabla v_n|^2 \geq \mu_2^{-1} S^2. \tag{4.69}$$

By (1.6) and Lemma 4.2, we conclude that

$$B_{\mu_1} + \frac{1}{4}\mu_2^{-1}S^2 \geq \lim_{n \to \infty} A_{\beta_n} = \frac{1}{4}(b_1 + b_2) \geq B > B_{\mu_1} + \frac{1}{4}\mu_2^{-1}S^2,$$

a contradiction. So Case 1 is impossible.

Case 2. Either $u_\infty \not\equiv 0$, $v_\infty \equiv 0$ or $u_\infty \equiv 0$, $v_\infty \not\equiv 0$.

First, we assume that $u_\infty \not\equiv 0$, $v_\infty \equiv 0$. Then (4.69) holds. Multiplying the equation of u in (4.1) by u_∞ and integrating over Ω, we obtain

$$\int_\Omega \nabla u_n \nabla u_\infty + \lambda_1 u_n u_\infty \leq \int_\Omega \mu_1 u_n^3 u_\infty.$$

Letting $n \to \infty$, we deduce from (4.26) that

$$\int_\Omega (|\nabla u_\infty|^2 + \lambda_1 u_\infty^2) \leq \int_\Omega \mu_1 u_\infty^4$$

$$\leq (4B_{\mu_1})^{-1} \left(\int_\Omega (|\nabla u_\infty|^2 + \lambda_1 u_\infty^2) \right)^2, \tag{4.70}$$

so

$$\int_\Omega (|\nabla u_\infty|^2 + \lambda_1 u_\infty^2) \geq 4B_{\mu_1}. \tag{4.71}$$

By (1.6) and Lemma 4.2, we easily conclude that

$$B_{\mu_1} + \frac{1}{4}\mu_2^{-1}S^2 \geq \lim_{n\to\infty} A_{\beta_n} = \lim_{n\to\infty} \frac{1}{4} \int_\Omega (|\nabla u_n|^2 + \lambda_1 u_n^2 + |\nabla v_n|^2 + \lambda_2 v_n^2)\, dx$$

$$= \frac{1}{4} \int_\Omega (|\nabla u_\infty|^2 + \lambda_1 u_\infty^2) + \frac{1}{4}b_2 + \lim_{n\to\infty} \frac{1}{4} \int_\Omega |\nabla(u_n - u_\infty)|^2$$

$$\geq B_{\mu_1} + \frac{1}{4}\mu_2^{-1}S^2 + \lim_{n\to\infty} \frac{1}{4} \int_\Omega |\nabla(u_n - u_\infty)|^2. \tag{4.72}$$

Thus, $\lim_{n\to\infty} \int_\Omega |\nabla(u_n - u_\infty)|^2 = 0$, namely $u_n \to u_\infty$ strongly in $H_0^1(\Omega)$. Moreover, $\int_\Omega (|\nabla u_\infty|^2 + \lambda_1 u_\infty^2) = 4B_{\mu_1}$, so we can derive from (4.70) and (4.71) that

$$\int_\Omega (|\nabla u_\infty|^2 + \lambda_1 u_\infty^2) = \int_\Omega \mu_1 u_\infty^4 = 4B_{\mu_1},$$

and then by [19] we see that u_∞ is a least energy solution of the Brezis–Nirenberg problem

$$-\Delta u + \lambda_1 u = \mu_1 u^3, \quad u \in H_0^1(\Omega).$$

The strong maximum principle gives $u_\infty > 0$ in Ω. Again by (4.72) we also have $b_2 = \mu_2^{-1}S^2$, which implies

$$\lim_{n\to\infty} \int_\Omega \beta_n u_n^2 v_n^2 = 0.$$

In fact, if $\lim_{n\to\infty} \int_\Omega \beta_n u_n^2 v_n^2 = -C < 0$, then

$$\int_\Omega (|\nabla v_n|^2 + \lambda_2 v_n^2) = \int_\Omega \mu_2 v_n^4 + \int_\Omega \beta_n u_n^2 v_n^2$$

$$\leq \mu_2 S^{-2} \left(\int_\Omega |\nabla v_n|^2 \right)^2 + \int_\Omega \beta_n u_n^2 v_n^2.$$

Letting $n \to \infty$, we obtain

$$b_2 \leq \mu_2 S^{-2} b_2^2 - C < \mu_2 S^{-2} b_2^2,$$

a contradiction. Thus, Theorem 4.4-(1) holds.

If $B_{\mu_1} + \frac{1}{4}\mu_2^{-1}S^2 > B_{\mu_2} + \frac{1}{4}\mu_1^{-1}S^2$, then (4.72) contradicts to Lemma 4.2, so Theorem 4.4-(1) can not hold, namely Remark 4.3 holds.

Now, we assume that $u_\infty \equiv 0$, $v_\infty \not\equiv 0$. Then by a similar argument, we see that (2) in Theorem 4.4 holds.

If both (1) and (2) in Theorem 4.4 do not hold, then we have $u_\infty \not\equiv 0$, $v_\infty \not\equiv 0$. Since $|\beta_n| \int_\Omega u_n^2 v_n^2 \leq \int_\Omega \mu_1 u_n^4 \leq C$, we deduce from Fatou Lemma that

$$\int_\Omega u_\infty^2 v_\infty^2 \leq \lim_{n\to\infty} \int_\Omega u_n^2 v_n^2 = 0. \tag{4.73}$$

Thus,

$$E_{\beta_n}\left(\sqrt{t}u_\infty, \sqrt{s}v_\infty\right) = \frac{1}{2}t \int_\Omega (|\nabla u_\infty|^2 + \lambda_1 u_\infty^2) - \frac{1}{4}t^2 \int_\Omega \mu_1 u_\infty^4$$

$$+ \frac{1}{2}s \int_\Omega (|\nabla v_\infty|^2 + \lambda_2 v_\infty^2) - \frac{1}{4}s^2 \int_\Omega \mu_2 v_\infty^4$$

$$= f(t) + g(s). \tag{4.74}$$

Similarly as Case 2, we see that (4.70) holds, so

$$\max_{t>0} f(t) = f(t_1) = \frac{1}{4}t_1 \int_\Omega (|\nabla u_\infty|^2 + \lambda_1 u_\infty^2),$$

where

$$t_1 = \frac{\int_\Omega (|\nabla u_\infty|^2 + \lambda_1 u_\infty^2)}{\int_\Omega \mu_1 u_\infty^4} \leq 1.$$

Similarly,

$$\max_{s>0} g(s) = g(s_1) = \frac{1}{4} s_1 \int_\Omega (|\nabla v_\infty|^2 + \lambda_2 v_\infty^2),$$

where

$$s_1 = \frac{\int_\Omega (|\nabla v_\infty|^2 + \lambda_2 v_\infty^2)}{\int_\Omega \mu_2 v_\infty^4} \leq 1.$$

So

$$\max_{t,s>0} E_{\beta_n} \left(\sqrt{t} u_\infty, \sqrt{s} v_\infty \right) = E_{\beta_n} \left(\sqrt{t_1} u_\infty, \sqrt{s_1} v_\infty \right) = f(t_1) + g(s_1),$$

and $(\sqrt{t_1} u_\infty, \sqrt{s_1} v_\infty) \in \mathcal{N}_{\beta_n}$. This implies $A_{\beta_n} \leq f(t_1) + g(s_1)$ and then

$$\lim_{n \to \infty} A_{\beta_n} \leq f(t_1) + g(s_1)$$

$$= \frac{1}{4} t_1 \int_\Omega (|\nabla u_\infty|^2 + \lambda_1 u_\infty^2) + \frac{1}{4} s_1 \int_\Omega (|\nabla v_\infty|^2 + \lambda_2 v_\infty^2)$$

$$\leq \frac{1}{4} \int_\Omega (|\nabla u_\infty|^2 + \lambda_1 u_\infty^2) + \frac{1}{4} \int_\Omega (|\nabla v_\infty|^2 + \lambda_2 v_\infty^2)$$

$$\leq \frac{1}{4} \int_\Omega (|\nabla u_\infty|^2 + \lambda_1 u_\infty^2) + \frac{1}{4} \int_\Omega (|\nabla v_\infty|^2 + \lambda_2 v_\infty^2)$$

$$+ \frac{1}{4} \lim_{n \to \infty} \int_\Omega (|\nabla (u_n - u_\infty)|^2 + |\nabla (v_n - v_\infty)|^2)$$

$$= \lim_{n \to \infty} \frac{1}{4} \int_\Omega (|\nabla u_n|^2 + \lambda_1 u_n^2 + |\nabla v_n|^2 + \lambda_2 v_n^2) \, dx$$

$$= \lim_{n \to \infty} A_{\beta_n}. \tag{4.75}$$

Therefore, $\lim_{n \to \infty} \int_\Omega (|\nabla (u_n - u_\infty)|^2 + |\nabla (v_n - v_\infty)|^2) = 0$, namely $(u_n, v_n) \to (u_\infty, v_\infty)$ strongly in H. Furthermore, $t_1 = s_1 = 1$. Since $\int_\Omega (|\nabla u_n|^2 + \lambda_1 u_n^2) = \int_\Omega (\mu_1 u_n^4 + \beta_n u_n^2 v_n^2)$, by letting $n \to \infty$ and using $t_1 = 1$, we also obtain

$$\lim_{n \to \infty} \int_\Omega \beta_n u_n^2 v_n^2 = 0.$$

Now we assume that u_∞ and v_∞ are both continuous (we will prove this later). Then we see from (4.73) that $u_\infty \cdot v_\infty \equiv 0$. Moreover $\{u_\infty > 0\}$ and $\{v_\infty > 0\}$ are disjoint and both open. By [19] we may let u_1 be a least energy solution of

$$-\Delta u + \lambda_1 u = \mu_1 u^3, \ u \in H_0^1(\{u_\infty > 0\}),$$

and v_1 be a least energy solution of

$$-\Delta v + \lambda_2 v = \mu_2 v^3, \ v \in H_0^1(\Omega \setminus \overline{\{u_\infty > 0\}}).$$

Since $t_1 = 1$, we have $E(u_1, 0) \le E(u_\infty, 0)$. Since $\{v_\infty > 0\} \subset \Omega \setminus \overline{\{u_\infty > 0\}}$, we also deduce from $s_1 = 1$ that $E(0, v_1) \le E(0, v_\infty)$. Noting that $(u_1, v_1) \in \mathcal{N}_{\beta_n}$, we see from (4.75) that

$$\lim_{n \to \infty} A_{\beta_n} \le E(u_1, v_1) = E(u_1, 0) + E(0, v_1)$$

$$\le E(u_\infty, 0) + E(0, v_\infty) = \lim_{n \to \infty} A_{\beta_n}. \tag{4.76}$$

Therefore, $E(u_1, 0) = E(u_\infty, 0)$ and then u_∞ is a positive least energy solution of

$$-\Delta u + \lambda_1 u = \mu_1 u^3, \ u \in H_0^1(\{u_\infty > 0\}). \tag{4.77}$$

Moreover, $E(0, v_1) = E(0, v_\infty)$ and so v_∞ is a least energy solution of

$$-\Delta v + \lambda_2 v = \mu_2 v^3, \ v \in H_0^1(\Omega \setminus \overline{\{u_\infty > 0\}}). \tag{4.78}$$

This also means that $\{u_\infty > 0\}$ is a connected domain. In fact, if $\{u_\infty > 0\}$ has at least two connected components Ω_1 and Ω_2, then we define

$$u_\infty^i = \chi_{\Omega_i} u_\infty, \quad i = 1, 2,$$

where

$$\chi_{\Omega_i}(x) = \begin{cases} 1, & x \in \Omega_i, \\ 0, & x \in \mathbb{R}^4 \setminus \Omega_i. \end{cases}$$

Then $u_\infty^i \in H_0^1(\{u_\infty > 0\}) \setminus \{0\}$. By $E'(u_\infty, 0)(u_\infty^i, 0) = 0$ we have

$$\int_{\{u_\infty > 0\}} |\nabla u_\infty^i|^2 + \lambda_1 |u_\infty^i|^2 \, dx = \int_{\{u_\infty > 0\}} \mu_1 |u_\infty^i|^4 \, dx, \quad i = 1, 2,$$

This implies $E(u_\infty, 0) \ge E(u_\infty^1, 0) + E(u_\infty^2, 0) \ge 2E(u_\infty, 0)$, a contradiction.

By a similar argument, we see that v_∞ is a positive least energy solution of

$$-\Delta v + \lambda_2 v = \mu_2 v^3, \ v \in H_0^1(\{v_\infty > 0\}),$$

$\{v_\infty > 0\}$ is connected, and u_∞ is a least energy solution of

$$-\Delta u + \lambda_1 u = \mu_1 u^3, \ u \in H_0^1(\Omega \setminus \overline{\{v_\infty > 0\}}).$$

It suffices to prove that

$$\{v_\infty > 0\} = \Omega \backslash \overline{\{u_\infty > 0\}}. \tag{4.79}$$

Noting that $\{v_\infty > 0\} \subset \Omega \backslash \overline{\{u_\infty > 0\}}$, we may assume that Ω_3 is the connected component of $\Omega \backslash \overline{\{u_\infty > 0\}}$ satisfying $\{v_\infty > 0\} \subset \Omega_3$. Then by a similar argument of (4.76)–(4.78), we can prove that v_∞ is a least energy solution of

$$-\Delta v + \lambda_2 v = \mu_2 v^3, \quad v \in H_0^1(\Omega_3).$$

By the strong maximum principle, we have $v_\infty > 0$ in Ω_3, namely $\{v_\infty > 0\} = \Omega_3$ and so $\{v_\infty > 0\}$ is a connected component of $\Omega \backslash \overline{\{u_\infty > 0\}}$. This implies that $\partial \{v_\infty > 0\} \subset \partial \Omega \cup \partial \{u_\infty > 0\}$ and so

$$\partial \{v_\infty > 0\} \backslash \partial \Omega \subset \partial \{u_\infty > 0\} \backslash \partial \Omega. \tag{4.80}$$

By a similar argument, we also have $\partial \{u_\infty > 0\} \backslash \partial \Omega \subset \partial \{v_\infty > 0\} \backslash \partial \Omega$, that is,

$$\partial \{v_\infty > 0\} \backslash \partial \Omega = \partial \{u_\infty > 0\} \backslash \partial \Omega.$$

Since $\{u_\infty > 0\}$ and $\{v_\infty > 0\}$ are both open domains, we conclude that (4.79) holds. Hence, (3) in Theorem 4.4 holds. Therefore, if we assume here the continuity of u_∞ and v_∞, the proof is complete. $\square$

Remark 4.5 Lemma 4.2 implies that

$$\lim_{\beta \to -\infty} A_\beta \le \min \left\{ B_{\mu_1} + \frac{1}{4} \mu_2^{-1} S^2, \ B_{\mu_2} + \frac{1}{4} \mu_1^{-1} S^2 \right\}.$$

If we have in addition that

$$\lim_{\beta \to -\infty} A_\beta < \min \left\{ B_{\mu_1} + \frac{1}{4} \mu_2^{-1} S^2, \ B_{\mu_2} + \frac{1}{4} \mu_1^{-1} S^2 \right\}, \tag{4.81}$$

then we can prove that only statement (3) in Theorem 4.4 holds. In fact, if (1) holds, we see from (4.72) that $\lim_{n \to \infty} A_{\beta_n} = B_{\mu_1} + \frac{1}{4} \mu_2^{-1} S^2$, a contradiction with (4.81). Similarly, if (2) holds, then $\lim_{n \to \infty} A_{\beta_n} = B_{\mu_2} + \frac{1}{4} \mu_1^{-1} S^2$, also a contradiction with (4.81). However, it seems very difficult for us to prove (4.81). Whether (4.81) hold or not remains an interesting open question.

From the previous proof, it suffices to prove that both u_∞ and v_∞ are continuous. First, we need to prove the L^∞-uniform bounds via the Moser iteration.

Lemma 4.5 *Let (u_n, v_n) be in Theorem 4.4 and assume that $(u_n, v_n) \to (u_\infty, v_\infty)$ strongly in H, then $\{u_n, v_n\}_n$ is uniformly bounded in $L^\infty(\Omega)$.*

Proof For any $s \geq 0$, first we claim that

$$\sup_n |u_n|_{2(s+1)} \leq C_1(s) \implies \sup_n |u_n|_{4(s+1)} \leq C_2(s), \tag{4.82}$$

where $C_i(s)(i = 1, 2)$ are positive constants independent of n.

Choose $l > 0$ and set

$$\psi_{n,l} := \min\{u_n^s, l\}, \quad \varphi_{n,l} = u_n \psi_{n,l}^2, \quad \Omega_{n,l} = \{x \in \Omega : u_n^s \leq l\},$$

$$\chi_{\Omega_{n,l}} = \begin{cases} 1 & \text{if } x \in \Omega_{n,l}, \\ 0 & \text{if } u \notin \Omega_{n,l}. \end{cases}$$

Then

$$\nabla(u_n \psi_{n,l}) = (1 + s\chi_{\Omega_{n,l}})\psi_{n,l}\nabla u_n, \quad \nabla\varphi_{n,l} = (1 + 2s\chi_{\Omega_{n,l}})\psi_{n,l}^2 \nabla u_n,$$

and $\varphi_{n,l} \in H_0^1(\Omega)$. Since

$$-\Delta u_n = |\lambda_1|u_n + \mu_1 u_n^3 + \beta_n u_n v_n^2 \leq |\lambda_1|u_n + \mu_1 u_n^3, \tag{4.83}$$

we have

$$\int_\Omega |\nabla u_n|^2 \psi_{n,l}^2 \leq \int_\Omega \nabla u_n \cdot \nabla\varphi_{n,l} \leq \int_\Omega (|\lambda_1|u_n + \mu_1 u_n^3)u_n \psi_{n,l}^2$$

$$\leq |\lambda_1| \int_\Omega u_n^{2(s+1)} + \mu_1 \int_\Omega u_n^4 \psi_{n,l}^2$$

$$\leq C + \mu_1 \int_\Omega u_n^4 \psi_{n,l}^2.$$

On the other hand, by Sobolev embedding (4.9) we have

$$\int_\Omega u_n^4 \psi_{n,l}^2 = \int_\Omega u_\infty^2 u_n^2 \psi_{n,l}^2 + \int_\Omega (u_n^2 - u_\infty^2)u_n^2 \psi_{n,l}^2$$

$$\leq k^2 \int_{\{u_\infty \leq k\}} u_n^{2(s+1)} + \int_{\{u_\infty > k\}} u_\infty^2 u_n^2 \psi_{n,l}^2$$

$$+ \left(\int_\Omega |u_n^2 - u_\infty^2|^2\right)^{1/2}\left(\int_\Omega u_n^4 \psi_{n,l}^4\right)^{1/2}$$

$$\leq C(k, n)\int_\Omega |\nabla(u_n\psi_{n,l})|^2 + Ck^2, \tag{4.84}$$

where

$$C(k, n) := S^{-1} \left(\int_{\Omega} |u_n^2 - u_\infty^2|^2 \right)^{1/2} + S^{-1} \left(\int_{\{u_\infty > k\}} u_\infty^4 \right)^{1/2}. \tag{4.85}$$

Therefore,

$$\int_{\Omega} |\nabla(u_n \psi_{n,l})|^2 \leq (1+s)^2 \int_{\Omega} |\nabla u_n|^2 \psi_{n,l}^2$$

$$\leq (1+s)^2 \mu_1 C(k, n) \int_{\Omega} |\nabla(u_n \psi_{n,l})|^2 + Ck^2 + C.$$

Since $u_n \to u_\infty$ in $H_0^1(\Omega)$, we have $u_n^2 \to u_\infty^2$ in $L^2(\Omega)$. By (4.85), there exists $k_0 > 0$ and $n_0 > 0$ large enough, such that for any $n \geq n_0$ we have $(1 + s)^2 \mu_1 C(k_0, n) \leq \frac{1}{2}$, where k_0 is independent of $n \in \mathbb{N}$. This implies that

$$\int_{\Omega_{n,l}} |\nabla(u_n^{s+1})|^2 \leq \int_{\Omega} |\nabla(u_n \psi_{n,l})|^2 \leq 2 C k_0^2 + 2C = C(s), \quad n \geq n_0.$$

Letting $l \to +\infty$, we get that $\int_{\Omega} |\nabla(u_n^{s+1})|^2 \leq C(s)$, $n \geq n_0$. Again by (4.9), we conclude that $\int_{\Omega} u_n^{4(s+1)} \leq S^{-2}C(s)^2$ for any $n \geq n_0$. On the other hand,

$$\int_{\Omega} u_n^4 \psi_{n,l}^2 \leq k^2 \int_{\{u_n \leq k\}} u_n^{2(s+1)} + \int_{\{u_n > k\}} u_n^2 u_n^2 \psi_{n,l}^2$$

$$\leq \widetilde{C}(k, n) \int_{\Omega} |\nabla(u_n \psi_{n,l})|^2 + Ck^2, \tag{4.86}$$

where

$$\widetilde{C}(k, n) := S^{-1} \left(\int_{\{u_n > k\}} u_n^4 \right)^{1/2}.$$

Since there exists $\tilde{k}_0 > 0$ large enough, such that $2(1+s)^2 \widetilde{C}(\tilde{k}_0, n) \leq \frac{1}{2}$ for any $n \leq n_0$, by repeating the arguments above, we have $\sup_{n \leq n_0} \int_{\Omega} u_n^{4(s+1)} \leq C$. This proves the claim.

Note that u_n is uniformly bounded in $H_0^1(\Omega)$ and so in $L^2(\Omega)$. Letting $s_1 = 0$ and using a bootstrap argument, we deduce from the claim that for any $q \geq 2$, there exists $C(q) > 0$ such that $\sup_n |u_n|_q \leq C(q)$. By (4.83) and [54, Theorem 8.17] we see that $\{u_n\}_n$ is uniformly bounded in $L^\infty(\Omega)$. Finally, a similar argument shows that $\{v_n\}_n$ is uniformly bounded in $L^\infty(\Omega)$. $\qquad\square$

We also need the following result from [73].

Theorem 4.6 (see [73]) *Let* $u_n, v_n \in H_0^1(\Omega)$ *be positive solutions of* (4.1) *with* $\beta = \beta_n$. *Assume that* $\{u_n, v_n\}_n$ *is uniformly bounded in* $L^\infty(\Omega)$. *Then for every* $\alpha \in (0, 1)$ *there exists* $C > 0$, *independent of* n, *such that*

$$\|(u_n, v_n)\|_{C^{0,\alpha}(\overline{\Omega})} \le C, \quad \text{for every } n \in \mathbb{N}.$$

Proof In [73], this theorem is proved for the subcritical case of dimension $N = 2, 3$. However, it is easy to see that the proof in [73] also works for the case $N = 4$, since all the integrals that appear in the proof are well defined, and the compactness of Sobolev embedding $H_0^1(\Omega) \hookrightarrow L^4(\Omega)$ is not used in the proof (personal communication with Hugo Tavares, an author of [73]). The reason for their only considering $N = 2, 3$ in [73] is following (personal communication with Hugo Tavares). In applications, first one obtains positive solutions of (4.1) with uniform E-bounds via variational methods (cf. [48, 87]). Second, the uniform E-bounds for a sequence of positive solutions of (4.1) (corresponding to unbounded $\beta \to -\infty$) yield uniform $H_0^1(\Omega)$-bounds. Third, uniform $H_0^1(\Omega)$-bounds implies uniform L^∞-bounds by standard elliptic regularity theories only if $N \le 3$, since one needs $3 < \frac{N+2}{N-2}$ for (4.1) with cubic nonlinearities. However, we can overcome this difficulty in the critical case $N = 4$ by Lemma 4.5. Therefore, we refer readers to [73] and omit the details of the proof for $N = 4$ here. $\qquad\square$

Now, we are in a position to finish the proof of Theorem 4.4.

Proof (Completion of the proof of (3) in Theorem 4.4) As pointed out before, it suffices to prove that u_∞ and v_∞ are continuous. Since $(u_n, v_n) \to (u_\infty, v_\infty)$ strongly in $H_0^1(\Omega) \times H_0^1(\Omega)$, by Lemma 4.5 we see that $\{u_n, v_n\}_n$ is uniformly bounded in $L^\infty(\Omega)$. Combining this with Theorem 4.6, we see that u_n, v_n are uniformly equicontinuous. Therefore, Ascoli-Arzelà theorem yields that u_∞ and v_∞ are both continuous. This completes the proof. $\qquad\square$

Remark 4.6 In subcritical case (i.e., $N \le 3$), one usually uses uniform $H_0^1(\Omega)$ bounds of positive solutions (u_β, v_β) to imply uniform $L^\infty(\Omega)$ bounds via a standard Morse iteration procedure. Then one can prove the convergence of (u_β, v_β) as $\beta \to -\infty$ with the help of the uniform L^∞ bounds. In the critical case $N = 4$, however, this idea can not be applied. Here, we use an opposite idea. That is, first we show the $H_0^1(\Omega)$ convergence of (u_β, v_β). Then we obtain the uniform L^∞ bounds by using the $H_0^1(\Omega)$ convergence (see Lemma 4.5). Once we have the uniform L^∞ bounds, we can deal with the problem in the critical case $N = 4$ just as in the subcritical case $N \le 3$ (see Theorem 4.6). Clearly, this idea also work for many other elliptic equations with critical growth.

Remark 4.7 After the publication of [33], we realized that, actually the continuity of u_∞ and v_∞ can be proved without using Lemma 4.5 and Theorem 4.6; see the next chapter for the new proof, where we will turn to study the more general critical system (1.8).

Chapter 5
A Generalized BEC System with Critical Exponents in Dimensions $N \geq 5$

Abstract As introduced in Chap. 1, we consider the generalized BEC system (1.8) with Sobolev critical exponents. In Chap. 4, we studied the special case $N = 4$, where system (1.8) is just the BEC system (1.2). In this chapter, we continue our previous study to consider the *higher dimensional* case $N \geq 5$. It turns out that some quite different phenomena appear comparing to the special case $N = 4$. For example, we can prove the existence of positive ground state solutions for *any* $\beta \neq 0$ (which can *not* hold in the special case $N = 4$). The *key* reason is $\frac{2^*}{2} = 2$ for $N = 4$ but $1 < \frac{2^*}{2} < 2$ for $N \geq 5$. The fact $1 < \frac{2^*}{2} < 2$ makes the problem quite different comparing to the case $\frac{2^*}{2} = 2$, and requires us to develop some different ideas and techniques.

5.1 Main Results

Consider the following generalized BEC system:

$$\begin{cases} -\Delta u + \lambda_1 u = \mu_1 u^{2^*-1} + \beta u^{\frac{2^*}{2}-1} v^{\frac{2^*}{2}}, & x \in \Omega, \\ -\Delta v + \lambda_2 v = \mu_2 v^{2^*-1} + \beta v^{\frac{2^*}{2}-1} u^{\frac{2^*}{2}}, & x \in \Omega, \\ u \geq 0, v \geq 0 \text{ in } \Omega, \quad u|_{\partial\Omega} = v|_{\partial\Omega} = 0, \end{cases} \tag{5.1}$$

where $\Omega \subset \mathbb{R}^N$ is a smooth-bounded domain and $-\lambda_1(\Omega) < \lambda_1, \lambda_2 < 0$. In Chap. 4, we studied the special case $N = 4$. In this chapter, we consider the higher dimensional case, namely we assume in this chapter that

$$N \geq 5 \quad \text{and denote } p := 2^*/2 \text{ for convenience.} \tag{5.2}$$

It turns out that some quite different phenomena happen comparing to the special case $N = 4$; see Remarks 5.1, 5.3, 5.5, and 5.6. The *key reason* is that $p = 2$ if $N = 4$ but $1 < p < 2$ when $N \geq 5$. The fact that $1 < p < 2$ makes the problem quite different (and also more complicated in proving some results such as Theorems 5.1 and 5.2) comparing with the case $p = 2$, and requires us to develop some different ideas and techniques.

© Springer-Verlag Berlin Heidelberg 2015
Z. Chen, *Solutions of Nonlinear Schrödinger Systems*, Springer Theses,
DOI 10.1007/978-3-662-45478-7_5

As pointed out before, system (5.1) can be seen as a critically coupled perturbed Brezis–Nirenberg problem. As we will see in Theorems 5.4 and 5.5, system (5.1) is closely related to the Brezis–Nirenberg problem (5.3). Using the same notations as in Chap. 4, we know that system (5.1) has two semi-trivial solutions $(u_{\mu_1}, 0)$ and $(0, u_{\mu_2})$, where u_{μ_i} is a positive least energy solution of the Brezis–Nirenberg problem

$$- \Delta u + \lambda_i u = \mu_i u^{2^*-1}, \quad u > 0 \text{ in } \Omega, \quad u|_{\partial\Omega} = 0. \tag{5.3}$$

Here, we are only concerned with nontrivial solutions of system (5.1). It is well known that nontrivial solutions of system (5.1) might be obtained by finding nontrivial critical points of a C^1 functional $E_\beta : H \to \mathbb{R}$, where E_β is defined by

$$E_\beta(u, v) = \frac{1}{2} \int_\Omega (|\nabla u|^2 + \lambda_1 u^2) + \frac{1}{2} \int_\Omega (|\nabla v|^2 + \lambda_2 v^2)$$
$$- \frac{1}{2p} \int_\Omega (\mu_1 |u|^{2p} + 2\beta |u|^p |v|^p + \mu_2 |v|^{2p}). \tag{5.4}$$

Define the Nehari manifold of system (5.1) as

$$\mathcal{N}_\beta = \left\{ (u, v) \in H \mid u \not\equiv 0, v \not\equiv 0, \int_\Omega (|\nabla u|^2 + \lambda_1 u^2) = \int_\Omega (\mu_1 |u|^{2p} + \beta |u|^p |v|^p), \right.$$
$$\left. \int_\Omega (|\nabla v|^2 + \lambda_2 v^2) = \int_\Omega (\mu_2 |v|^{2p} + \beta |u|^p |v|^p) \right\}.$$

Clearly, all nontrivial solutions of system (5.1) belong to $\mathcal{N}_\beta$. Similarly as in Chap. 1, it is trivial to see $\mathcal{N}_\beta \neq \emptyset$. Define

$$A_\beta := \inf_{(u,v)\in\mathcal{N}_\beta} E_\beta(u, v) = \inf_{(u,v)\in\mathcal{N}_\beta} \frac{1}{N} \int_\Omega (|\nabla u|^2 + \lambda_1 u^2 + |\nabla v|^2 + \lambda_2 v^2) \, dx. \tag{5.5}$$

By the Sobolev inequality, we know that $A_\beta > 0$, for all $\beta \in \mathbb{R}$. Define as in Chap. 1 that

Definition 5.1 We call that a solution (u, v) of system (5.1) is a *ground state* solution (or a *least energy* solution), if (u, v) is nontrivial and $E_\beta(u, v) = A_\beta$. We call that the least energy A_β is attained, if there exists $(u, v) \in \mathcal{N}_\beta$ such that $E_\beta(u, v) = A_\beta$.

For convenience as before, we will omit the subscript β when there is no confusion arising. As in Chap. 4, first we consider the symmetric case $-\lambda_1(\Omega) < \lambda_1 = \lambda_2 = \lambda < 0$. By Brezis and Nirenberg [19], we know that the Brezis–Nirenberg problem

$$- \Delta u + \lambda u = u^{2^*-1}, \quad u > 0 \text{ in } \Omega, \quad u|_{\partial\Omega} = 0 \tag{5.6}$$

has a positive least energy solution ω with least energy

$$B_1 := \frac{1}{N} \int_\Omega (|\nabla \omega|^2 + \lambda \omega^2) \, dx = \frac{1}{N} \int_\Omega \omega^{2^*} \, dx. \tag{5.7}$$

Moreover,

$$\int_\Omega (|\nabla u|^2 + \lambda u^2) \, dx \geq (NB_1)^{2/N} \left(\int_\Omega |u|^{2^*} \, dx \right)^{2/2^*}, \quad \forall u \in H_0^1(\Omega). \tag{5.8}$$

consider the following nonlinear problem ($p = \frac{N}{N-2} < 2$ since $N \geq 5$)

$$\begin{cases} \mu_1 k^{p-1} + \beta k^{p/2-1} l^{p/2} = 1, \\ \beta k^{p/2} l^{p/2-1} + \mu_2 l^{p-1} = 1, \\ k > 0, \ l > 0. \end{cases} \tag{5.9}$$

We will prove in Lemma 5.1 that there exists (k_0, l_0) such that

$$(k_0, l_0) \text{ satisfies } (5.9) \text{ and } k_0 = \min \{k : (k, l) \text{ is a solution of } (5.9)\}. \tag{5.10}$$

Our first result is

Theorem 5.1 *Assume that* $-\lambda_1(\Omega) < \lambda_1 = \lambda_2 = \lambda < 0$. *Let* (k_0, l_0) *in* (5.10). *Then for any* $\beta > 0$, $(\sqrt{k_0}\omega, \sqrt{l_0}\omega)$ *is a positive solution of* (5.1). *Moreover, if* $\beta \geq \frac{2}{N-2} \max\{\mu_1, \mu_2\}$, *then* $E(\sqrt{k_0}\omega, \sqrt{l_0}\omega) = A$, *namely* $(\sqrt{k_0}\omega, \sqrt{l_0}\omega)$ *is a positive ground state solution of system* (5.1).

Remark 5.1 (1) In the special case $N = 4$ and $2p = 2^*$, Theorem 4.1 in Chap. 4 indicates that system (5.1) has no nontrivial nonnegative solutions if $\beta \in [\min\{\mu_1, \mu_2\}, \max\{\mu_1, \mu_2\}]$ and $\mu_1 \neq \mu_2$ (the reason is $p = 2$). Therefore, the general case $N \geq 5$ is quite different from the case $N = 4$. As we will see in Sect. 5.2 that, the proof of Theorem 5.1 is much more delicate than Chap. 4 due to the nonlinearity of problem (5.9) (clearly (5.9) is a trivial *linear* system if $N = 4$).

(2) Similarly as in Remark 4.2 of Chap. 4, it is easy to prove that, if Ω is star-shaped, the assumption $-\lambda_1(\Omega) < \lambda < 0$ in Theorem 5.1 is also optimal.

Our following result deals with the classification of the ground state solutions.

Theorem 5.2 *Let assumptions in Theorem 5.1 hold. Then there exists a constant* $\beta_0 \geq \frac{2}{N-2} \max\{\mu_1, \mu_2\}$ *determined only by* (μ_1, μ_2), *such that, if* $\beta > \beta_0$ *and* (u, v) *is any a positive ground state solution of* (5.1), *then* $(u, v) = (\sqrt{k_0}\omega, \sqrt{l_0}\omega)$, *where* ω *is a positive least energy solution of the Brezis–Nirenberg problem* (5.6). *In particular, the positive ground state solution of* (5.1) *is unique if* $\Omega \subset \mathbb{R}^N$ *is a ball.*

Remark 5.2 We can give a precise definition of β_0 (see (5.106) in Sect. 5.4). In particular, $\beta_0 = \frac{2}{N-2} \max\{\mu_1, \mu_2\}$ if $\mu_1 = \mu_2$. The basic idea of proving Theorem 5.2 is the same as Theorem 4.2. However, some different techniques are also needed due to the nonlinearity of system (5.9).

Now let us consider the general case $-\lambda_1(\Omega) < \lambda_1, \lambda_2 < 0$. The following result is interesting to ourselves.

Theorem 5.3 *Assume that* $-\lambda_1(\Omega) < \lambda_1, \lambda_2 < 0$. *Then system* (5.1) *has a positive ground state solution* (u, v) *with* $E(u, v) = A$ *for any* $\beta \neq 0$.

Remark 5.3 For the general case $\lambda_1 \leq \lambda_2$, when $N = 4$ and $2p = 2^*$, Theorem 4.3 in Chap. 4 indicates that (5.1) has a positive ground state solution for any

$$\beta \in (-\infty, 0) \cup (0, \beta_1) \cup (\beta_2, +\infty),$$

where $\beta_i, i = 1, 2$ are some positive constants satisfying

$$\beta_1 \leq \min\{\mu_1, \mu_2\} \leq \max\{\mu_1, \mu_2\} \leq \beta_2.$$

That is, we do not know whether the ground state solution exists or not if $\beta \in [\beta_1, \beta_2]$ (in the symmetric case $\lambda_1 = \lambda_2$, Theorem 4.1 shows that nontrivial positive solutions do not exist if $\beta \in [\min\{\mu_1, \mu_2\}, \max\{\mu_1, \mu_2\}]$ and $\mu_1 \neq \mu_2$). Comparing this with Theorem 5.3, it turns out that the general case $N \geq 5$ is quite different from the special case $N = 4$. As we will see in the proof, the *essential* reason is also $1 < p < 2$ for $N \geq 5$.

Remark 5.4 Recently, Kim [58] also studied system (5.1) for all $N \geq 3$. He proved the existence of $\beta_0 > 0$ such that, system (5.1) has positive ground state solutions for any $\beta > \beta_0$. For $|\beta|$ sufficiently small, the existence of nontrivial solutions was also proved in [58] via a perturbation approach, but the solutions obtained there seem not necessary to be ground state solutions. Clearly, for the case $N \geq 4$, Theorems 4.3 and 5.3 are much more general than those results in [58]. We also remark that any other results in Chaps. 4 and 5 (such as the classification and asymptotic behaviors of ground state solutions) can not be found in [58].

Now we study the asymptotic behaviors of the ground state solutions in the repulsive case $\beta \to -\infty$ (compare with Theorem 4.4).

Theorem 5.4 *Assume that* $-\lambda_1(\Omega) < \lambda_1, \lambda_2 < 0$. *Let* $\beta_n < 0$, $n \in \mathbb{N}$, *satisfy* $\beta_n \to -\infty$ *as* $n \to \infty$, *and* (u_n, v_n) *be the positive ground state solutions of* (5.1) *with* $\beta = \beta_n$. *Then* $\int_\Omega \beta_n u_n^p v_n^p \, dx \to 0$ *as* $n \to \infty$, *and passing to a subsequence, one of the following conclusions holds.*

(1) $u_n \to u_\infty$ *strongly in* $H_0^1(\Omega)$ *and* $v_n \rightharpoonup 0$ *weakly in* $H_0^1(\Omega)$ *(so* $v_n \to 0$ *for almost every* $x \in \Omega$), *where* u_∞ *is a positive least energy solution of*

$$-\Delta u + \lambda_1 u = \mu_1 |u|^{2^*-2} u, \quad u \in H_0^1(\Omega).$$

(2) $v_n \to v_\infty$ strongly in $H_0^1(\Omega)$ and $u_n \rightharpoonup 0$ weakly in $H_0^1(\Omega)$ (so $u_n \to 0$ for almost every $x \in \Omega$), where v_∞ is a positive least energy solution of

$$-\Delta v + \lambda_2 v = \mu_2 |v|^{2^*-2} v, \quad v \in H_0^1(\Omega).$$

(3) $(u_n, v_n) \to (u_\infty, v_\infty)$ strongly in H and $u_\infty \cdot v_\infty \equiv 0$, where $u_\infty \in C(\overline{\Omega})$ is a positive least energy solution of

$$-\Delta u + \lambda_1 u = \mu_1 |u|^{2^*-2} u, \quad u \in H_0^1(\{u_\infty > 0\}),$$

and $v_\infty \in C(\overline{\Omega})$ is a positive least energy solution of

$$-\Delta v + \lambda_2 v = \mu_2 |v|^{2^*-2} v, \quad v \in H_0^1(\{v_\infty > 0\}).$$

Furthermore, both $\{v_\infty > 0\}$ and $\{u_\infty > 0\}$ are connected domains, and $\{v_\infty > 0\} = \Omega \backslash \overline{\{u_\infty > 0\}}$.

In particular, if $N \geq 6$, then only conclusion (3) holds, and $u_\infty - v_\infty$ is a least energy sign-changing solution to problem

$$- \Delta u + \lambda_1 u^+ - \lambda_2 u^- = \mu_1 (u^+)^{2^*-1} - \mu_2 (u^-)^{2^*-1}, \quad u \in H_0^1(\Omega). \quad (5.11)$$

Here, a sign-changing solution u of (5.11) is called a *least energy sign-changing solution*, if u attains the minimal functional energy among all sign-changing solutions of (5.11). Remark that, (5.11) can be seen as a generalized Brezis–Nirenberg problem, since (5.11) is just the classical Brezis–Nirenberg problem provided $\lambda_1 = \lambda_2$ and $\mu_1 = \mu_2$. As an application of Theorem 5.4, we turn to consider the Brezis–Nirenberg problem

$$- \Delta u + \lambda_1 u = \mu_1 |u|^{2^*-2} u, \quad u \in H_0^1(\Omega), \quad (5.12)$$

where $-\lambda_1(\Omega) < \lambda_1 < 0$. Its corresponding functional is

$$J(u) := \frac{1}{2} \int_\Omega (|\nabla u|^2 + \lambda_1 u^2)\, dx - \frac{1}{2^*} \int_\Omega \mu_1 |u|^{2^*}\, dx, \quad u \in H_0^1(\Omega).$$

Then, we have the following result.

Theorem 5.5 *Assume $N \geq 6$. Let (u_∞, v_∞) be in Theorem 5.4 in the symmetric case where $\lambda_2 = \lambda_1$ and $\mu_2 = \mu_1$. Then $u_\infty - v_\infty$ is a least energy sign-changing solution of (5.12), and*

$$J(u_\infty - v_\infty) < B_{\mu_1} + \frac{1}{N} \mu_1^{\frac{2-N}{2}} S^{\frac{N}{2}}, \quad (5.13)$$

where S is seen in (5.15) and B_{μ_1} is the least energy of problem (5.12) (see (5.59) in Sect. 5.3).

Remark 5.5 (i) Theorem 5.4 has been proved for the special case $N = 4$ and $2p = 2^*$ in Theorem 4.4, where we raised an open question: Can one show that only conclusion (3) holds? As pointed out before, the results in Chap. 4 have been published in our recent paper [33]. In particular, the referee of [33] pointed out that this question may be related to the existence of sign-changing solutions to the Brezis–Nirenberg problem. Motivated by the referee's comment, it is natural for us to consider (5.1) under assumption (5.2) in this chapter. Here in the case $N \geq 6$, with the help of a sharp energy estimate result from [32], we exclude statements (1)–(2) and verify the referee's comment successfully. In particular, this result indicates that system (5.1) is really closely related to the Brezis–Nirenberg problem. Unfortunately, we do not know whether only Theorem 5.4-(3) holds for $N = 4, 5$, which still remains as an interesting open question.

(ii)   In the proof of Theorem 5.4-(3), a key point is to prove the continuity of u_∞ and v_∞. We remark here that, our proof of the continuity of u_∞ and v_∞ is completely different from that of Thoerem 4.4 for the special case $N = 4$, and can also be used to the special case $N = 4$.

(iii)   The existence of least energy sign-changing solutions to the Brezis–Nirenberg problem (5.12) in the case $N \geq 6$ was proved by Cerami et al. [24] in 1986. Here, Theorem 5.5 is a direct corollary of Theorem 5.4, and so the proof of Theorem 5.5 is completely different from [24].

Similarly as in Chap. 4, to prove Theorem 5.3, we also have to study the existence of ground state solutions to the corresponding limit problem

$$\begin{cases} -\Delta u = \mu_1 |u|^{2p-2}u + \beta|u|^{p-2}u|v|^p, & x \in \mathbb{R}^N, \\ -\Delta v = \mu_2 |v|^{2p-2}v + \beta|v|^{p-2}v|u|^p, & x \in \mathbb{R}^N, \\ u, v \in D^{1,2}(\mathbb{R}^N). \end{cases} \tag{5.14}$$

Here $D^{1,2}(\mathbb{R}^N) := \{u \in L^{2^*}(\mathbb{R}^N) \,|\, |\nabla u| \in L^2(\mathbb{R}^N)\}$ with the standard norm $\|u\|_{D^{1,2}} := (\int_{\mathbb{R}^N} |\nabla u|^2 \, dx)^{1/2}$. Let S be the sharp constant of the Sobolev embedding $D^{1,2}(\mathbb{R}^N) \hookrightarrow L^{2^*}(\mathbb{R}^N)$

$$\int_{\mathbb{R}^N} |\nabla u|^2 \, dx \geq S \left(\int_{\mathbb{R}^N} |u|^{2^*} \, dx \right)^{\frac{2}{2^*}}. \tag{5.15}$$

For any $\varepsilon > 0$ and $y \in \mathbb{R}^N$, we define $U_{\varepsilon,y} \in D^{1,2}(\mathbb{R}^N)$ by

$$U_{\varepsilon,y}(x) := [N(N-2)]^{\frac{N-2}{4}} \left(\frac{\varepsilon}{\varepsilon^2 + |x - y|^2} \right)^{\frac{N-2}{2}}. \tag{5.16}$$

Then $U_{\varepsilon,y}$ satisfies $-\Delta u = |u|^{2^*-2}u$ in $\mathbb{R}^N$, and

$$\int_{\mathbb{R}^N} |\nabla U_{\varepsilon,y}|^2 \, dx = \int_{\mathbb{R}^N} |U_{\varepsilon,y}|^{2^*} \, dx = S^{N/2}. \tag{5.17}$$

Moreover, $\{U_{\varepsilon,y} : \varepsilon > 0, y \in \mathbb{R}^N\}$ contains all positive solutions of equation $-\Delta u = |u|^{2^*-2}u$ in $\mathbb{R}^N$. See [12, 82] for details.

Also system (5.14) has two semi-trivial solutions $(\mu_1^{\frac{2-N}{4}} U_{\varepsilon,y}, 0)$ and $(0, \mu_2^{\frac{2-N}{4}} U_{\varepsilon,y})$. To obtain nontrival solutions of (5.14), we denote $D := D^{1,2}(\mathbb{R}^N) \times D^{1,2}(\mathbb{R}^N)$ and define a C^1 functional $I : D \to \mathbb{R}$ by

$$I(u, v) = \frac{1}{2} \int_{\mathbb{R}^N} (|\nabla u|^2 + |\nabla v|^2) - \frac{1}{2p} \int_{\mathbb{R}^N} (\mu_1|u|^{2p} + 2\beta|u|^p|v|^p + \mu_2|v|^{2p}). \tag{5.18}$$

Define the Nehari manifold of system (5.14) as

$$\mathcal{M} = \left\{ (u, v) \in D \,\bigg|\, u \not\equiv 0, v \not\equiv 0, \int_{\mathbb{R}^N} |\nabla u|^2 = \int_{\mathbb{R}^N} (\mu_1|u|^{2p} + \beta|u|^p|v|^p), \right.$$

$$\left. \int_{\mathbb{R}^N} |\nabla v|^2 = \int_{\mathbb{R}^N} (\mu_2|v|^{2p} + \beta|u|^p|v|^p) \right\}.$$

Then all nontrivial solutions of system (5.14) belong to $\mathcal{M}$. Similarly as $\mathcal{N}$, we have $\mathcal{M} \neq \emptyset$. Define

$$B := \inf_{(u,v)\in\mathcal{M}} I(u, v) = \inf_{(u,v)\in\mathcal{M}} \frac{1}{N} \int_{\mathbb{R}^N} \left(|\nabla u|^2 + |\nabla v|^2 \right) dx. \tag{5.19}$$

Then we have the following result, which plays a crucial role in the proof of Theorem 5.3.

Theorem 5.6 (1) *If $\beta < 0$, then B is not attained.*

(2) *If $\beta > 0$, then (5.14) has a positive ground state solution (U, V) with $I(U, V) = B$, which is radially symmetric decreasing. Moreover,*

(2-1) *if $\beta \geq \frac{2}{N-2} \max\{\mu_1, \mu_2\}$, then $I(\sqrt{k_0}U_{\varepsilon,y}, \sqrt{l_0}U_{\varepsilon,y}) = B$, where (k_0, l_0) in (5.10). That is, $(\sqrt{k_0}U_{\varepsilon,y}, \sqrt{l_0}U_{\varepsilon,y})$ is a positive ground state solution of (5.14).*

(2-2) *there exists $0 < \beta_1 \leq \frac{2}{N-2} \max\{\mu_1, \mu_2\}$, and for any $0 < \beta < \beta_1$, there exists a solution $(k(\beta), l(\beta))$ of (5.9), such that*

$$I\left(\sqrt{k(\beta)}U_{\varepsilon,y}, \sqrt{l(\beta)}U_{\varepsilon,y}\right) > B = I(U, V).$$

That is, $(\sqrt{k(\beta)}U_{\varepsilon,y}, \sqrt{l(\beta)}U_{\varepsilon,y})$ is a different positive solution of (5.14) with respect to (U, V).

Remark 5.6 For the special case where $N = 4$ and $2p = 2^*$, Theorem 4.5 in Chap. 4 indicates that system (5.14) has no nontrivial positive solutions if $\beta \in [\min\{\mu_1, \mu_2\}, \max\{\mu_1, \mu_2\}]$ and $\mu_1 \neq \mu_2$; $(\sqrt{k(\beta)}U_{\varepsilon,y}, \sqrt{l(\beta)}U_{\varepsilon,y})$ is a positive ground state solution of (5.14) if $0 < \beta < \min\{\mu_1, \mu_2\}$. Hence, the general case $N \geq 5$ is quite different from the case $N = 4$. As we will see in Sect. 5.2, the idea of proving Theorem 5.6-(2) in case $0 < \beta < \frac{2}{N-2}\max\{\mu_1, \mu_2\}$, which also works for the case $\beta \geq \frac{2}{N-2}\max\{\mu_1, \mu_2\}$, is completely different from that of the speical case $N = 4$ in Chap. 4.

Remark 5.7 The above results in this chapter, which were published in a joint work with Zou [42], are all concerned with ground state solutions of system (5.1). Clearly, ground state solutions are all *positive*. In a subsequent work [25], among other things, we studied the blow-up behavior of the ground state solutions as $\lambda_1, \lambda_2 \to 0$. It turns out that, after a suitable scaling, the limit profile is just the Green function. In another subsequent work [29], we studied the existence and asymptotic behavior of *semi-nodal* solutions (i.e., one component of the solution changing sign and the other one positive; see Chap. 3 for definition) for system (5.1) in the case where $N \geq 6$ and $\beta < 0$. Whether (5.1) has *sign-changing* solutions or not remains open.

We will prove these theorems in the rest of this chapter, and the basic ideas of the following proofs are similar to those in Chap. 4. However, as pointed out in Remarks 5.1, 5.3, 5.5, and 5.6, the general case $N \geq 5$ (i.e., $1 < p < 2$) is quite different from $N = 4$ (i.e., $p = 2$), and so some different ideas will be developed. As before, we denote the norm of $L^p(\Omega)$ by $|u|_p = (\int_\Omega |u|^p \, dx)^{\frac{1}{p}}$, and the norm of $H_0^1(\Omega)$ by $\|u\| = |\nabla u|_2$.

5.2 The Symmetric Case $\lambda_1 = \lambda_2$

Define functions

$$\alpha_1(k, l) := \mu_1 k^{p-1} + \beta k^{\frac{p}{2}-1}l^{\frac{p}{2}} - 1, \quad k > 0, \ l \geq 0; \tag{5.20}$$

$$\alpha_2(k, l) := \mu_2 l^{p-1} + \beta l^{\frac{p}{2}-1}k^{\frac{p}{2}} - 1, \quad l > 0, \ k \geq 0; \tag{5.21}$$

$$h_1(k) := \beta^{-2/p}\left(k^{1-p/2} - \mu_1 k^{p/2}\right)^{2/p}, \quad 0 < k \leq \mu_1^{-\frac{1}{p-1}}; \tag{5.22}$$

$$h_2(l) := \beta^{-2/p}\left(l^{1-p/2} - \mu_2 l^{p/2}\right)^{2/p}, \quad 0 < l \leq \mu_2^{-\frac{1}{p-1}}. \tag{5.23}$$

Clearly, $\alpha_1(k, h_1(k)) \equiv 0$ and $\alpha_2(h_2(l), l) \equiv 0$.

Lemma 5.1 *Assume that $\beta > 0$, then equation*

$$\alpha_1(k, l) = 0, \quad \alpha_2(k, l) = 0, \quad k, l > 0 \tag{5.24}$$

has a solution (k_0, l_0), which satisfies

$$\alpha_2(k, h_1(k)) < 0, \quad \forall 0 < k < k_0, \tag{5.25}$$

namely (k_0, l_0) satisfies (5.10). Similarly, (5.24) has a solution (k_1, l_1) such that

$$\alpha_1(h_2(l), l) < 0, \quad \forall 0 < l < l_1. \tag{5.26}$$

Proof Equation $\alpha_1(k, l) = 0$, $k, l > 0$ imply that

$$l = h_1(k), \quad 0 < k < \mu_1^{-\frac{1}{p-1}}.$$

On the other hand, $\alpha_2(k, l) = 0$ implies that $\mu_2 l^{p/2} + \beta k^{p/2} = l^{1-p/2}$. Therefore, we turn to show that

$$\mu_2 \frac{1 - \mu_1 k^{p-1}}{\beta k^{p/2-1}} + \beta k^{p/2} = \left(\frac{1 - \mu_1 k^{p-1}}{\beta k^{p/2-1}} \right)^{\frac{2-p}{p}}, \quad 0 < k^{p-1} < \frac{1}{\mu_1} \tag{5.27}$$

has a solution. Obviously, (5.27) is equivalent to

$$f(k) := \left(\frac{1}{\beta k^{p-1}} - \frac{\mu_1}{\beta} \right)^{\frac{2-p}{p}} - \frac{\mu_2}{\beta} - \frac{\beta^2 - \mu_1 \mu_2}{\beta} k^{p-1} = 0, \quad 0 < k^{p-1} < \frac{1}{\mu_1}. \tag{5.28}$$

Recalling that $N \geq 5$ and $2p = 2^*$, we have $2 - p > 0$ and so

$$\lim_{k \to 0+} f(k) = +\infty, \quad f(\mu_1^{-\frac{1}{p-1}}) = -\frac{\beta}{\mu_1} < 0.$$

Therefore, there exists $k_0 \in (0, \mu_1^{-\frac{1}{p-1}})$ such that $f(k_0) = 0$ and $f(k) > 0$ for $k \in (0, k_0)$. Let $l_0 = h_1(k_0)$, then (k_0, l_0) is a solution of (5.24). Moreover, (5.25) follows directly from $f(k) > 0$ for $k \in (0, k_0)$. The existence of (k_1, l_1) that satisfy (5.24) and (5.26) is similar. □

Lemma 5.2 *Assume that $\beta \geq (p - 1) \max\{\mu_1, \mu_2\}$, then $h_1(k) + k$ is strictly increasing for $k \in [0, \mu_1^{-\frac{1}{p-1}}]$ and $h_2(l) + l$ is strictly increasing for $l \in [0, \mu_2^{-\frac{1}{p-1}}]$.*

Proof Since for $k > 0$

$$h'_1(k) = \frac{2}{p}\beta^{-2/p}\left(k^{1-p/2} - \mu_1 k^{p/2}\right)^{2/p-1}\left((1-p/2)k^{-p/2} - \frac{p}{2}\mu_1 k^{p/2-1}\right),$$

we see that $h'_1(k) \geq 0$ for $0 < \mu_1 k^{p-1} \leq \frac{2-p}{p}$ or $\mu_1 k^{p-1} = 1$, and $h'_1(k) < 0$ for $\frac{2-p}{p} < \mu_1 k^{p-1} < 1$. By direct computations, we can derive from $h''_1(k) = 0$ and $\frac{2-p}{p} < \mu_1 k^{p-1} < 1$ that $k = (\mu_1 p)^{-\frac{1}{p-1}}$. This, together with $\beta \geq (p-1)\max\{\mu_1, \mu_2\}$, gives

$$\min_{0 < k^{p-1} \leq \mu_1^{-1}} h'_1(k) = h'_1\left((\mu_1 p)^{-\frac{1}{p-1}}\right) = -\beta^{-2/p}\mu_1^{2/p}(p-1)^{2/p} \geq -1,$$

and so $h'_1(k) > -1$ for any $0 < k \leq \mu_1^{-\frac{1}{p-1}}$ with $k \neq (\mu_1 p)^{-\frac{1}{p-1}}$. This implies that $h_1(k) + k$ is strictly increasing for $k \in [0, \mu_1^{-\frac{1}{p-1}}]$. Similarly, $h_2(l) + l$ is strictly increasing for $l \in [0, \mu_2^{-\frac{1}{p-1}}]$. $\square$

Lemma 5.3 *Assume that $\beta \geq (p-1)\max\{\mu_1, \mu_2\}$. Let (k_0, l_0) be in Lemma 5.1. Then $\max\{\mu_1(k_0 + l_0)^{p-1}, \mu_2(k_0 + l_0)^{p-1}\} < 1$ and*

$$\alpha_2(k, h_1(k)) < 0, \quad \forall 0 < k < k_0; \quad \alpha_1(h_2(l), l) < 0, \quad \forall 0 < l < l_0. \quad (5.29)$$

Proof By Lemma 5.2 we have

$$h_1(\mu_1^{-\frac{1}{p-1}}) + \mu_1^{-\frac{1}{p-1}} = \mu_1^{-\frac{1}{p-1}} > h_1(k_0) + k_0 = k_0 + l_0,$$

namely $\mu_1(k_0 + l_0)^{p-1} < 1$. Similarly, $\mu_2(k_0 + l_0)^{p-1} < 1$. By Lemma 5.1, to prove (5.29), it suffices to prove that $(k_0, l_0) = (k_1, l_1)$. By (5.25)–(5.26), we see that $k_1 \geq k_0$, $l_0 \geq l_1$. If $k_1 > k_0$, then $k_1 + h_1(k_1) > k_0 + h_1(k_0)$, that is, $h_2(l_1) + l_1 = k_1 + l_1 > k_0 + l_0 = h_2(l_0) + l_0$, and so $l_1 > l_0$, a contradiction. Therefore, $k_1 = k_0$ and $l_0 = l_1$. This completes the proof. $\square$

Lemma 5.4 *Assume that $\beta \geq (p-1)\max\{\mu_1, \mu_2\}$. Then*

$$\begin{cases} k + l \leq k_0 + l_0, \\ \alpha_1(k, l) \geq 0, \quad \alpha_2(k, l) \geq 0, \\ k, l \geq 0, \quad (k, l) \neq (0, 0) \end{cases} \quad (5.30)$$

has a unique solution (k_0, l_0).

Proof Clearly, (k_0, l_0) satisfies (5.30). Let $(\tilde{k}, \tilde{l})$ be any a solution of (5.30). Without loss of generality, we assume that $\tilde{k} > 0$. If $\tilde{l} = 0$, then by $\tilde{k} \leq k_0 + l_0$ and $\alpha_1(\tilde{k}, 0) \geq 0$ we get that

$$1 \leq \mu_1 \tilde{k}^{p-1} \leq \mu_1 (k_0 + l_0)^{p-1},$$

which contradicts with Lemma 5.3. Therefore, $\tilde{l} > 0$.

Assume by contradiction that $\tilde{k} < k_0$. Similarly to the proof of Lemma 5.2, by (5.23) it is easy to see that $h_2(l)$ is strictly increasing for $0 < \mu_2 l^{p-1} \leq \frac{2-p}{p}$, and strictly decreasing for $\frac{2-p}{p} \leq \mu_2 l^{p-1} \leq 1$. Moreover, $h_2(0) = h_2(\mu_2^{-\frac{1}{p-1}}) = 0$. Since $0 < \tilde{k} < k_0 = h_2(l_0)$, there exists $0 < l_2 < l_3 < \mu_2^{-\frac{1}{p-1}}$ such that $h_2(l_2) = h_2(l_3) = \tilde{k}$ and

$$\alpha_2(\tilde{k}, l) < 0 \iff h_2(l) > \tilde{k} \iff l_2 < l < l_3. \tag{5.31}$$

Since $\alpha_2(\tilde{k}, \tilde{l}) \geq 0$, we have $\tilde{l} \leq l_2$ or $\tilde{l} \geq l_3$. Since $\alpha_1(\tilde{k}, \tilde{l}) \geq 0$, we have $\tilde{l} \geq h_1(\tilde{k})$. By Lemma 5.3, we have $\alpha_2(\tilde{k}, h_1(\tilde{k})) < 0$, and so $l_2 < h_1(\tilde{k}) < l_3$. These imply

$$\tilde{l} \geq l_3. \tag{5.32}$$

On the other hand, since $l_1 := k_0 + l_0 - \tilde{k} > l_0$, we have

$$h_2(l_1) + k_0 + l_0 - \tilde{k} = h_2(l_1) + l_1 > h_2(l_0) + l_0 = k_0 + l_0,$$

that is, $h_2(l_1) > \tilde{k}$. By (5.31), we have $l_2 < l_1 < l_3$. By $\tilde{k} + \tilde{l} \leq k_0 + l_0$, we have

$$\tilde{l} \leq l_1 < l_3,$$

which contradicts with (5.32). Therefore, $\tilde{k} \geq k_0$. A similar argument also shows $\tilde{l} \geq l_0$. Therefore, $(\tilde{k}, \tilde{l}) = (k_0, l_0)$. □

Now we can give the proof of Theorem 5.1.

Proof (Proof of Theorem 5.1) Assume that $-\lambda_1(\Omega) < \lambda_1 = \lambda_2 = \lambda < 0$. As before, we have $A > 0$. Since $\beta > 0$, Lemma 5.1 indicates that Eq. (5.9) has a solution (k_0, l_0). Recalling (5.7), we see that $(\sqrt{k_0}\omega, \sqrt{l_0}\omega)$ is a nontrivial solution of (5.1) and

$$0 < A \leq E(\sqrt{k_0}\omega, \sqrt{l_0}\omega) = (k_0 + l_0)B_1. \tag{5.33}$$

Now we assume that $\beta \geq (p-1)\max\{\mu_1, \mu_2\}$, and we want to show that $A = E(\sqrt{k_0}\omega, \sqrt{l_0}\omega)$. Let $\{(u_n, v_n)\} \subset \mathcal{N}$ be a minimizing sequence for A, namely $E(u_n, v_n) \to A$. Define

$$c_n = \left(\int_\Omega |u_n|^{2p} \, dx \right)^{1/p}, \quad d_n = \left(\int_\Omega |v_n|^{2p} \, dx \right)^{1/p}.$$

By (5.8) we have

$$(NB_1)^{2/N} c_n \leq \int_\Omega (|\nabla u_n|^2 + \lambda u_n^2) = \int_\Omega (\mu_1 |u_n|^{2p} + \beta |u_n|^p |v_n|^p)$$

$$\leq \mu_1 c_n^p + \beta c_n^{p/2} d_n^{p/2}, \tag{5.34}$$

$$(NB_1)^{2/N} d_n \leq \int_\Omega (|\nabla v_n|^2 + \lambda v_n^2) = \int_\Omega (\mu_2 |v_n|^{2p} + \beta |u_n|^p |v_n|^p)$$

$$\leq \mu_2 d_n^p + \beta c_n^{p/2} d_n^{p/2}. \tag{5.35}$$

Since $E(u_n, v_n) = \frac{1}{N} \int_\Omega (|\nabla u_n|^2 + \lambda u_n^2 + |\nabla v_n|^2 + \lambda v_n^2)$, by (5.33) we have

$$(NB_1)^{2/N} (c_n + d_n) \leq NE(u_n, v_n) \leq N(k_0 + l_0)B_1 + o(1), \tag{5.36}$$

$$\mu_1 c_n^{p-1} + \beta c_n^{p/2-1} d_n^{p/2} \geq (NB_1)^{2/N}, \tag{5.37}$$

$$\mu_2 d_n^{p-1} + \beta c_n^{p/2} d_n^{p/2-1} \geq (NB_1)^{2/N}. \tag{5.38}$$

First, this means that c_n and d_n are both uniformly bounded. Passing to a subsequence, we assume that $c_n \to c$ and $d_n \to d$. Then by (5.34)–(5.35), we have $\mu_1 c^p + 2\beta c^{p/2} d^{p/2} + \mu_2 d^p \geq NA > 0$. Hence, without loss of generality, we assume that $c > 0$. If $d = 0$, then (5.36) implies $c \leq (NB_1)^{1-2/N} (k_0 + l_0)$. By (5.37) and Lemma 5.3, we get

$$(NB_1)^{2/N} \leq \mu_1 c^{p-1} \leq \mu_1 (k_0 + l_0)^{p-1} (NB_1)^{2/N} < (NB_1)^{2/N},$$

a contradiction. Therefore, $c > 0$ and $d > 0$. Let $k = \frac{c}{(NB_1)^{1-2/N}}$ and $l = \frac{d}{(NB_1)^{1-2/N}}$, then by (5.36)–(5.38) we see that (k, l) satisfies (5.30). Consequently, Lemma 5.4 gives $(k, l) = (k_0, l_0)$. It follows that $c_n \to k_0 (NB_1)^{1-2/N}$ and $d_n \to l_0 (NB_1)^{1-2/N}$ as $n \to +\infty$, and

$$NA = \lim_{n \to +\infty} NE(u_n, v_n) \geq \lim_{n \to +\infty} (NB_1)^{2/N} (c_n + d_n) = N(k_0 + l_0)B_1.$$

Combining this with (5.33), we conclude

$$A = (k_0 + l_0)B_1 = E(\sqrt{k_0}\omega, \sqrt{l_0}\omega),$$

and so $(\sqrt{k_0}\omega, \sqrt{l_0}\omega)$ is a positive ground state solution of system (5.1). $\quad\square$

In the rest of this section, we study the limit problem (5.14) and give the proof of Theorem 5.6. Differently from Theorem 4.5 in Chap. 4, the proof of Theorem 5.6 is quite long, and we divide it into several parts. First, by the Sobolev inequality (5.15) we have

$$B = \inf_{(u,v)\in\mathcal{N}} \frac{1}{4} \int_{\mathbb{R}^N} \left(|\nabla u|^2 + |\nabla v|^2 \right) dx > 0. \tag{5.39}$$

The following lemma is the counterpart of Lemma 4.1. Remark that Lemma 4.1 holds for any $\beta < \sqrt{\mu_1\mu_2}$, but we can only prove the validity of the following lemma for $\beta < 0$.

Lemma 5.5 *If A (resp. B) is attained by a couple $(u, v) \in \mathcal{N}$ (resp. $(u, v) \in \mathcal{M}$), then this couple is a critical point of E (resp. I), provided $-\infty < \beta < 0$.*

Proof Let $\beta < 0$. Assume that $(u, v) \in \mathcal{M}$ such that $B = I(u, v)$. Define

$$G_1(u, v) := I'(u, v)(u, 0) = \int_{\mathbb{R}^N} |\nabla u|^2 - \int_{\mathbb{R}^N} (\mu_1|u|^{2p} + \beta|u|^p|v|^p),$$

$$G_2(u, v) := I'(u, v)(0, v) = \int_{\mathbb{R}^N} |\nabla v|^2 - \int_{\mathbb{R}^N} (\mu_2|v|^{2p} + \beta|u|^p|v|^p).$$

Then there exists two Lagrange multipliers $L_1, L_2 \in \mathbb{R}$ such that

$$I'(u, v) + L_1 G_1'(u, v) + L_2 G_2'(u, v) = 0.$$

Testing this equation with $(u, 0)$ and $(0, v)$, respectively, we obtain

$$\left((2p-2) \int_{\mathbb{R}^N} \mu_1|u|^{2p} - (2-p) \int_{\mathbb{R}^N} \beta|u|^p|v|^p \right) L_1 + L_2 p \int_{\mathbb{R}^N} \beta|u|^p|v|^p = 0,$$

$$\left((2p-2) \int_{\mathbb{R}^N} \mu_2|v|^{2p} - (2-p) \int_{\mathbb{R}^N} \beta|u|^p|v|^p \right) L_2 + L_1 p \int_{\mathbb{R}^N} \beta|u|^p|v|^p = 0.$$

Since $\beta < 0$, we deduce from $G_1(u, v) = G_2(u, v) = 0$ that

$$\left((2p-2) \int_{\mathbb{R}^N} \mu_1|u|^{2p} - (2-p) \int_{\mathbb{R}^N} \beta|u|^p|v|^p \right)$$

$$\times \left((2p-2) \int_{\mathbb{R}^N} \mu_2|v|^{2p} - (2-p) \int_{\mathbb{R}^N} \beta|u|^p|v|^p \right) > \left(p \int_{\mathbb{R}^N} \beta|u|^p|v|^p \right)^2.$$

From this, we derive $L_1 = L_2 = 0$ and so $I'(u, v) = 0$. Similarly, if $(u, v) \in \mathcal{N}$ such that $E(u, v) = A$, then $E'(u, v) = 0$. □

First, we can prove Theorem 5.6 for the repulsive case $\beta < 0$.

Proof (Proof of Theorem 5.6 (1)) This proof is similar to the proof of Theorem 4.5 in case $N = 4$, but the details are more delicate. By (5.16), we see that $\omega_{\mu_i} := \mu_i^{\frac{2-N}{4}} U_{1,0}$ satisfies equation $-\Delta u = \mu_i |u|^{2^*-2} u$ in $\mathbb{R}^N$. Let $e_1 = (1, 0, \ldots, 0) \in \mathbb{R}^N$ and

$$(u_R(x), v_R(x)) = (\omega_{\mu_1}(x), \omega_{\mu_2}(x + Re_1)).$$

Then $v_R \rightharpoonup 0$ weakly in $D^{1,2}(\mathbb{R}^N)$ and so $v_R^p \rightharpoonup 0$ weakly in $L^2(\mathbb{R}^N)$ as $R \to +\infty$. Consequently,

$$\lim_{R \to +\infty} \int_{\mathbb{R}^N} u_R^p v_R^p \, dx = 0.$$

Note that $\beta < 0$. For $R > 0$ sufficiently large, by a similar argument as that of Lemma 5.1 (or see the argument of existing $(t_\varepsilon, s_\varepsilon)$ in the proof of Lemma 5.8), we conclude that

$$\begin{cases} t^2 \int_{\mathbb{R}^N} |\nabla u_R|^2 \, dx = t^2 \mu_1 \int_{\mathbb{R}^N} u_R^{2p} \, dx = t^{2p} \mu_1 \int_{\mathbb{R}^N} u_R^{2p} \, dx + t^p s^p \beta \int_{\mathbb{R}^N} u_R^p v_R^p \, dx, \\ s^2 \int_{\mathbb{R}^N} |\nabla v_R|^2 \, dx = s^2 \mu_2 \int_{\mathbb{R}^N} v_R^{2p} \, dx = s^{2p} \mu_2 \int_{\mathbb{R}^N} v_R^{2p} \, dx + t^p s^p \beta \int_{\mathbb{R}^N} u_R^p v_R^p \, dx, \end{cases}$$

have a solution (t_R, s_R) with $t_R > 1$ and $s_R > 1$. For simplicity, we denote

$$D_1 := \mu_1 \int_{\mathbb{R}^N} u_R^{2p} \, dx = \mu_1 \int_{\mathbb{R}^N} \omega_{\mu_1}^{2p} \, dx > 0,$$

$$D_2 := \mu_2 \int_{\mathbb{R}^N} v_R^{2p} \, dx = \mu_2 \int_{\mathbb{R}^N} \omega_{\mu_2}^{2p} \, dx > 0,$$

$$F_R := |\beta| \int_{\mathbb{R}^N} u_R^p v_R^p \, dx \to 0, \quad \text{as } R \to +\infty.$$

Then

$$t_R^2 D_1 = t_R^{2p} D_1 - t_R^p s_R^p F_R, \quad s_R^2 D_2 = s_R^{2p} D_2 - t_R^p s_R^p F_R. \tag{5.40}$$

Assume that, up to a subsequence, $t_R \to +\infty$ as $R \to \infty$, then by

$$t_R^{2p} D_1 - t_R^2 D_1 = s_R^{2p} D_2 - s_R^2 D_2$$

we also have $s_R \to +\infty$. Since $2 - p < p$, we have

$$t_R^p D_1 - t_R^{2-p} D_1 \geq \frac{1}{2} t_R^p D_1, \quad s_R^p D_2 - s_R^{2-p} D_2 \geq \frac{1}{2} s_R^p D_2, \quad \text{for } R \text{ large enough,}$$

and so

$$F_R = \frac{t_R^p - t_R^{2-p}}{s_R^p} D_1 \geq \frac{t_R^p}{2s_R^p} D_1, \quad F_R = \frac{s_R^p - s_R^{2-p}}{t_R^p} D_2 \geq \frac{s_R^p}{2t_R^p} D_2,$$

which implies that

$$0 < \frac{1}{4} D_1 D_2 \leq F_R^2 \to 0, \quad \text{as } R \to +\infty,$$

, a contradiction. Therefore, t_R and s_R are uniformly bounded. Then by (5.40) and $F_R \to 0$ as $R \to \infty$, we conclude that

$$\lim_{R \to +\infty} (|t_R - 1| + |s_R - 1|) = 0.$$

Noting that $(t_R u_R, s_R v_R) \in \mathcal{M}$, it follows from (5.17) that

$$B \leq I(t_R u_R, s_R v_R) = \frac{1}{N} \left(t_R^2 \int_{\mathbb{R}^N} |\nabla u_R|^2 \, dx + s_R^2 \int_{\mathbb{R}^N} |\nabla v_R|^2 \, dx \right)$$

$$= \frac{1}{N} \left(t_R^2 \mu_1^{\frac{2-N}{2}} + s_R^2 \mu_2^{\frac{2-N}{2}} \right) S^{N/2}.$$

Letting $R \to +\infty$, we get that $B \leq \frac{1}{N} (\mu_1^{\frac{2-N}{2}} + \mu_2^{\frac{2-N}{2}}) S^{N/2}$.

On the other hand, by a similar argument as Theorem 4.5-(1), we easily obtain $B \geq \frac{1}{N} (\mu_1^{\frac{2-N}{2}} + \mu_2^{\frac{2-N}{2}}) S^{N/2}$. Hence,

$$B = \frac{1}{N} \left(\mu_1^{\frac{2-N}{2}} + \mu_2^{\frac{2-N}{2}} \right) S^{N/2}. \tag{5.41}$$

Now, assume by contradiction that B is attained by some $(u, v) \in \mathcal{M}$. Then by repeating the argument of Theorem 4.5-(1) with minor modifications, we can obtain

$$B = I(u, v) = \frac{1}{N} \int_{\mathbb{R}^N} (|\nabla u|^2 + |\nabla v|^2) \, dx > \frac{1}{N} \left(\mu_1^{\frac{2-N}{2}} + \mu_2^{\frac{2-N}{2}} \right) S^{N/2},$$

which is a contradiction. This completes the proof. $\qquad \square$

Proof (Proof of Theorem 5.6 (2-1)) This proof is similar to the proof of Theorem 5.1. Since $\beta > 0$, $(\sqrt{k_0}U_{\varepsilon,y}, \sqrt{l_0}U_{\varepsilon,y})$ is a nontrivial solution of (5.14) and

$$B \leq I\left(\sqrt{k_0}U_{\varepsilon,y}, \sqrt{l_0}U_{\varepsilon,y}\right) = \frac{1}{N}(k_0 + l_0)S^{N/2}. \tag{5.42}$$

Assume that $\beta \geq (p-1)\max\{\mu_1, \mu_2\}$. Let $\{(u_n, v_n)\} \subset \mathcal{M}$ be a minimizing sequence for B, namely $I(u_n, v_n) \to B$. Define $c_n = \left(\int_{\mathbb{R}^N} |u_n|^{2p}\, dx\right)^{1/p}$ and $d_n = \left(\int_{\mathbb{R}^N} |v_n|^{2p}\, dx\right)^{1/p}$ as before, we have

$$Sc_n \leq \int_{\mathbb{R}^N} |\nabla u_n|^2\, dx = \int_{\mathbb{R}^N} \mu_1 |u_n|^{2p} + \beta |u_n|^p |v_n|^p\, dx \leq \mu_1 c_n^p + \beta c_n^{p/2} d_n^{p/2},$$

$$Sd_n \leq \int_{\mathbb{R}^N} |\nabla v_n|^2\, dx = \int_{\mathbb{R}^N} \mu_2 |v_n|^{2p} + \beta |u_n|^p |v_n|^p\, dx \leq \mu_2 d_n^p + \beta c_n^{p/2} d_n^{p/2}.$$

This implies

$$S(c_n + d_n) \leq NI(u_n, v_n) \leq (k_0 + l_0)S^{N/2} + o(1),$$
$$\mu_1 c_n^{p-1} + \beta c_n^{p/2-1} d_n^{p/2} \geq S, \quad \beta c_n^{p/2} d_n^{p/2-1} + \mu_2 d_n^{p-1} \geq S.$$

Similarly as in the proof of Theorem 5.1, we conclude that $c_n \to k_0 S^{N/2-1}$ and $d_n \to l_0 S^{N/2-1}$ as $n \to +\infty$, and so

$$NB = \lim_{n \to +\infty} NI(u_n, v_n) \geq \lim_{n \to +\infty} S(c_n + d_n) = (k_0 + l_0)S^{N/2}.$$

This implies that

$$B = \frac{1}{N}(k_0 + l_0)S^{N/2} = I(\sqrt{k_0}U_{\varepsilon,y}, \sqrt{l_0}U_{\varepsilon,y}), \tag{5.43}$$

and so $(\sqrt{k_0}U_{\varepsilon,y}, \sqrt{l_0}U_{\varepsilon,y})$ is a positive ground state solution of (5.14). $\square$

To finish the proof of Theorem 5.6, we need to show that (5.14) has a positive ground state solution for any $0 < \beta < (p-1)\max\{\mu_1, \mu_2\}$. The following proof works for all $\beta > 0$. Therefore, we assume that $\beta > 0$, and define

$$B' := \inf_{(u,v) \in \mathcal{M}'} I(u, v), \tag{5.44}$$

where

$$\mathcal{M}' := \left\{(u, v) \in D \setminus \{(0,0)\} \mid I'(u, v)(u, v) = 0\right\}. \tag{5.45}$$

Clearly $B' \leq B$ since $\mathcal{M} \subset \mathcal{M}'$. By Sobolev inequality, we have $B' > 0$. Define $B(0, R) := \{x \in \mathbb{R}^N : |x| < R\}$ and $H(0, R) := H_0^1(B(0, R)) \times H_0^1(B(0, R))$. Consider

$$\begin{cases} -\Delta u = \mu_1 |u|^{2p-2} u + \beta |u|^{p-2} u |v|^p, & x \in B(0, R), \\ -\Delta v = \mu_2 |v|^{2p-2} v + \beta |v|^{p-2} v |u|^p, & x \in B(0, R), \\ u, v \in H_0^1(B(0, R)), \end{cases} \tag{5.46}$$

and define

$$B'(R) := \inf_{(u,v) \in \mathcal{M}'(R)} I(u, v), \tag{5.47}$$

where

$$\mathcal{M}'(R) := \left\{ (u, v) \in H(0, R) \setminus \{(0, 0)\} \, \Big| \, \int_{B(0,R)} (|\nabla u|^2 + |\nabla v|^2) \right.$$

$$\left. - \int_{B(0,R)} (\mu_1 |u|^{2p} + 2\beta |u|^p |v|^p + \mu_2 |v|^{2p}) = 0 \right\}. \tag{5.48}$$

Lemma 5.6 $B'(R) \equiv B'$ for all $R > 0$.

Proof Take any $R_1 > R_2$. By $\mathcal{M}'(R_2) \subset \mathcal{M}'(R_1)$, we have $B'(R_1) \leq B'(R_2)$. On the other hand, for any $(u, v) \in \mathcal{M}'(R_1)$, we define

$$(u_1(x), v_1(x)) := \left(\left(\frac{R_1}{R_2}\right)^{\frac{N-2}{2}} u\left(\frac{R_1}{R_2} x\right), \left(\frac{R_1}{R_2}\right)^{\frac{N-2}{2}} v\left(\frac{R_1}{R_2} x\right) \right),$$

then it is easy to check that $(u_1, v_1) \in \mathcal{M}'(R_2)$, so

$$B'(R_2) \leq I(u_1, v_1) = I(u, v), \quad \forall (u, v) \in \mathcal{M}'(R_1).$$

That is, $B'(R_2) \leq B'(R_1)$ and so $B'(R_1) = B'(R_2)$.

Clearly, $B' \leq B'(R)$. Let $(u_n, v_n) \in \mathcal{M}'$ be a minimizing sequence of B'. Moreover, we may assume that $u_n, v_n \in H_0^1(B(0, R_n))$ for some $R_n > 0$. Then $(u_n, v_n) \in \mathcal{M}'(R_n)$ and

$$B' = \lim_{n \to \infty} I(u_n, v_n) \geq \lim_{n \to \infty} B'(R_n) \equiv B'(R).$$

Therefore, $B'(R) \equiv B'$ for all $R > 0$. $\qquad\qquad\qquad\qquad\qquad\qquad\qquad \square$

Let $0 \leq \varepsilon < p - 1$. Consider

$$\begin{cases} -\Delta u = \mu_1 |u|^{2p-2-2\varepsilon} u + \beta |u|^{p-2-\varepsilon} u |v|^{p-\varepsilon}, & x \in B(0, 1), \\ -\Delta v = \mu_2 |v|^{2p-2-2\varepsilon} v + \beta |v|^{p-2-\varepsilon} v |u|^{p-\varepsilon}, & x \in B(0, 1), \\ u, v \in H_0^1(B(0, 1)), \end{cases} \tag{5.49}$$

and define

$$B_\varepsilon := \inf_{(u,v) \in \mathcal{M}_\varepsilon'} I_\varepsilon(u, v), \tag{5.50}$$

where

$$I_\varepsilon(u, v) := \frac{1}{2} \int_{B(0,1)} (|\nabla u|^2 + |\nabla v|^2)$$
$$- \frac{1}{2p - 2\varepsilon} \int_{B(0,1)} (\mu_1 |u|^{2p-2\varepsilon} + 2\beta |u|^{p-\varepsilon} |v|^{p-\varepsilon} + \mu_2 |v|^{2p-2\varepsilon}),$$

$$\mathcal{M}_\varepsilon' := \left\{ (u, v) \in H(0, 1) \setminus \{(0, 0)\}, \ H_\varepsilon(u, v) := I_\varepsilon'(u, v)(u, v) = 0 \right\}.$$

Lemma 5.7 *For any $0 < \varepsilon < p - 1$, there holds*

$$B_\varepsilon < \min \left\{ \inf_{(u,0) \in \mathcal{M}_\varepsilon'} I_\varepsilon(u, 0), \ \inf_{(0,v) \in \mathcal{M}_\varepsilon'} I_\varepsilon(0, v) \right\}.$$

Proof Fix any $0 < \varepsilon < p - 1$. Recalling $2 < 2p - 2\varepsilon < 2^*$, we may let u_i be a least energy solution of

$$-\Delta u = \mu_i |u|^{2p-2-2\varepsilon} u, \quad u \in H_0^1(B(0, 1)).$$

Then

$$I_\varepsilon(u_1, 0) = c_1 := \inf_{(u,0) \in \mathcal{M}_\varepsilon'} I_\varepsilon(u, 0), \quad I_\varepsilon(0, u_2) = c_2 := \inf_{(0,v) \in \mathcal{M}_\varepsilon'} I_\varepsilon(0, v).$$

The following proof is inspired by Abdellaoui et al. [1]. For any $s \in \mathbb{R}$, there exists a unique $t(s) > 0$ such that $(t(s)u_1, t(s)su_2) \in \mathcal{M}_\varepsilon'$. In fact,

$$t(s)^{2p-2\varepsilon-2} = \frac{\int_{B(0,1)} (|\nabla u_1|^2 + s^2 |\nabla u_2|^2)}{\int_{B(0,1)} (\mu_1 |u_1|^{2p-2\varepsilon} + 2\beta |u_1|^{p-\varepsilon} |su_2|^{p-\varepsilon} + \mu_2 |su_2|^{2p-2\varepsilon})}$$
$$= \frac{p'c_1 + s^2 p'c_2}{p'c_1 + |s|^{2p-2\varepsilon} p'c_2 + |s|^{p-\varepsilon} \int_{B(0,1)} 2\beta |u_1|^{p-\varepsilon} |u_2|^{p-\varepsilon}},$$

where $p' = \frac{2p-2\varepsilon}{p-1-\varepsilon}$. We note that $t(0) = 1$. Since $1 < p - \varepsilon < 2$, by direct computations we have

$$\lim_{s\to 0} \frac{t'(s)}{|s|^{p-2-\varepsilon}s} = -\frac{(p-\varepsilon)\int_{B(0,1)} 2\beta |u_1|^{p-\varepsilon}|u_2|^{p-\varepsilon}}{(2p-2\varepsilon-2)p'c_1},$$

that is,

$$t'(s) = -\frac{(p-\varepsilon)\int_{B(0,1)} 2\beta |u_1|^{p-\varepsilon}|u_2|^{p-\varepsilon}}{(2p-2\varepsilon-2)p'c_1}|s|^{p-\varepsilon-2}s(1+o(1)), \quad \text{as } s \to 0,$$

and so

$$t(s) = 1 - \frac{\int_{B(0,1)} 2\beta |u_1|^{p-\varepsilon}|u_2|^{p-\varepsilon}}{(2p-2\varepsilon-2)p'c_1}|s|^{p-\varepsilon}(1+o(1)), \quad \text{as } s \to 0.$$

This implies that

$$t(s)^{2p-2\varepsilon} = 1 - \frac{(2p-2\varepsilon)\int_{B(0,1)} 2\beta |u_1|^{p-\varepsilon}|u_2|^{p-\varepsilon}}{(2p-2\varepsilon-2)p'c_1}|s|^{p-\varepsilon}(1+o(1))$$

$$= 1 - \frac{\int_{B(0,1)} 2\beta |u_1|^{p-\varepsilon}|u_2|^{p-\varepsilon}}{2c_1}|s|^{p-\varepsilon}(1+o(1)), \quad \text{as } s \to 0.$$

Therefore, we derive from $1/2 - 1/p' = 1/(2p-2\varepsilon) > 0$ that

$$B_\varepsilon \leq I_\varepsilon\left(t(s)u_1, t(s)su_2\right)$$

$$= \frac{t(s)^{2p-2\varepsilon}}{p'}\left(p'c_1 + |s|^{2p-2\varepsilon}p'c_2 + |s|^{p-\varepsilon}\int_{B(0,1)} 2\beta |u_1|^{p-\varepsilon}|u_2|^{p-\varepsilon}\right)$$

$$= c_1 - \left(\frac{1}{2} - \frac{1}{p'}\right)|s|^{p-\varepsilon}\int_{B(0,1)} 2\beta |u_1|^{p-\varepsilon}|u_2|^{p-\varepsilon} + o(|s|^{p-\varepsilon})$$

$$< c_1 = \inf_{(u,0)\in\mathcal{M}'_\varepsilon} I_\varepsilon(u,0) \quad \text{as } |s| > 0 \text{ small enough.}$$

A similar argument also shows $B_\varepsilon < \inf_{(0,v)\in\mathcal{M}'_\varepsilon} I_\varepsilon(0,v)$. □

Recalling ω_{μ_i} in the proof of Theorem 5.6-(1), similarly as Lemma 5.7, we have

$$B' < \min\left\{\inf_{(u,0)\in\mathcal{M}'} I(u,0), \inf_{(0,v)\in\mathcal{M}'} I(0,v)\right\} = \min\left\{I(\omega_{\mu_1},0), I(0,\omega_{\mu_2})\right\}$$

$$= \min\left\{\frac{1}{N}\mu_2^{\frac{2-N}{2}}S^{N/2}, \frac{1}{N}\mu_1^{\frac{2-N}{2}}S^{N/2}\right\}. \tag{5.51}$$

Theorem 5.7 *For any* $0 < \varepsilon < p - 1$, (5.49) *has a classical ground state solution* $(u_\varepsilon, v_\varepsilon)$, *and* $u_\varepsilon, v_\varepsilon$ *are both positive radially symmetric decreasing.*

Proof Fix any $0 < \varepsilon < p - 1$, it is easy to see that $B_\varepsilon > 0$. For $(u, v) \in \mathscr{M}'_\varepsilon$ with $u \geq 0, v \geq 0$, we denote by (u^*, v^*) as its Schwartz symmetrization. Then by the properties of Schwartz symmetrization and $\beta > 0$, we have

$$\int_{B(0,1)} (|\nabla u^*|^2 + |\nabla v^*|^2) \leq \int_{B(0,1)} (\mu_1 |u^*|^{2p-2\varepsilon} + 2\beta |u^*|^{p-\varepsilon} |v^*|^{p-\varepsilon} + \mu_2 |v^*|^{2p-2\varepsilon}).$$

Consequently, there exists $0 < t^* \leq 1$ such that $(t^* u^*, t^* v^*) \in \mathscr{M}'_\varepsilon$, and then

$$I_\varepsilon(t^* u^*, t^* v^*) = \left(\frac{1}{2} - \frac{1}{2p - 2\varepsilon} \right) (t^*)^2 \int_{B(0,1)} (|\nabla u^*|^2 + |\nabla v^*|^2)$$

$$\leq \left(\frac{1}{2} - \frac{1}{2p - 2\varepsilon} \right) \int_{B(0,1)} (|\nabla u|^2 + |\nabla v|^2) = I_\varepsilon(u, v). \quad (5.52)$$

Therefore, we may take a minimizing sequence $(u_n, v_n) \in \mathscr{M}'_\varepsilon$ of B_ε such that $(u_n, v_n) = (u_n^*, v_n^*)$ and $I_\varepsilon(u_n, v_n) \to B_\varepsilon$. It follows from (5.52) that u_n, v_n are uniformly bounded in $H_0^1(B(0, 1))$. Passing to a subsequence, we may assume that $u_n \rightharpoonup u_\varepsilon, v_n \rightharpoonup v_\varepsilon$ weakly in $H_0^1(B(0, 1))$. By the compactness of the embedding $H_0^1(B(0, 1)) \hookrightarrow L^{2p-2\varepsilon}(B(0, 1))$, we have

$$\int_{B(0,1)} (\mu_1 |u_\varepsilon|^{2p-2\varepsilon} + 2\beta |u_\varepsilon|^{p-\varepsilon} |v_\varepsilon|^{p-\varepsilon} + \mu_2 |v_\varepsilon|^{2p-2\varepsilon})$$

$$= \lim_{n \to \infty} \int_{B(0,1)} (\mu_1 |u_n|^{2p-2\varepsilon} + 2\beta |u_n|^{p-\varepsilon} |v_n|^{p-\varepsilon} + \mu_2 |v_n|^{2p-2\varepsilon})$$

$$= \frac{2p - 2\varepsilon}{p - 1 - \varepsilon} \lim_{n \to \infty} I_\varepsilon(u_n, v_n) = \frac{2p - 2\varepsilon}{p - 1 - \varepsilon} B_\varepsilon > 0,$$

which implies $(u_\varepsilon, v_\varepsilon) \neq (0, 0)$. Moreover, $u_\varepsilon \geq 0, v_\varepsilon \geq 0$ are radially symmetric. Meanwhile, $\int_{B(0,1)} (|\nabla u_\varepsilon|^2 + |\nabla v_\varepsilon|^2) \leq \lim_{n \to \infty} \int_{B(0,1)} (|\nabla u_n|^2 + |\nabla v_n|^2)$, so

$$\int_{B(0,1)} (|\nabla u_\varepsilon|^2 + |\nabla v_\varepsilon|^2) \leq \int_{B(0,1)} (\mu_1 u_\varepsilon^{2p-2\varepsilon} + 2\beta u_\varepsilon^{p-\varepsilon} v_\varepsilon^{p-\varepsilon} + \mu_2 v_\varepsilon^{2p-2\varepsilon}).$$

Consequently, there exists $0 < t_\varepsilon \leq 1$ such that $(t_\varepsilon u_\varepsilon, t_\varepsilon v_\varepsilon) \in \mathscr{M}'_\varepsilon$, and then

$$
\begin{aligned}
B_\varepsilon &\leq I_\varepsilon(t_\varepsilon u_\varepsilon, t_\varepsilon v_\varepsilon) = \left(\frac{1}{2} - \frac{1}{2p - 2\varepsilon}\right)(t_\varepsilon)^2 \int_{B(0,1)} (|\nabla u_\varepsilon|^2 + |\nabla v_\varepsilon|^2) \\
&\leq \lim_{n\to\infty} \left(\frac{1}{2} - \frac{1}{2p - 2\varepsilon}\right) \int_{B(0,1)} (|\nabla u_n|^2 + |\nabla v_n|^2) \\
&= \lim_{n\to\infty} I_\varepsilon(u_n, v_n) = B_\varepsilon.
\end{aligned}
$$

Therefore, $t_\varepsilon = 1$ and $(u_\varepsilon, v_\varepsilon) \in \mathscr{M}'_\varepsilon$ with $I(u_\varepsilon, v_\varepsilon) = B_\varepsilon$. Moreover,

$$
\int_{B(0,1)} (|\nabla u_\varepsilon|^2 + |\nabla v_\varepsilon|^2) = \lim_{n\to\infty} \int_{B(0,1)} (|\nabla u_n|^2 + |\nabla v_n|^2),
$$

namely $u_n \to u_\varepsilon$ and $v_n \to v_\varepsilon$ strongly in $H_0^1(B(0, 1))$. On the other hand, there exists a Lagrange multiplier $\gamma \in \mathbb{R}$ such that

$$
I'_\varepsilon(u_\varepsilon, v_\varepsilon) - \gamma H'_\varepsilon(u_\varepsilon, v_\varepsilon) = 0.
$$

Since $I'_\varepsilon(u_\varepsilon, v_\varepsilon)(u_\varepsilon, v_\varepsilon) = H_\varepsilon(u_\varepsilon, v_\varepsilon) = 0$ and

$$
\begin{aligned}
&H'_\varepsilon(u_\varepsilon, v_\varepsilon)(u_\varepsilon, v_\varepsilon) \\
&= (2 + 2\varepsilon - 2p) \int_{B(0,1)} (\mu_1 u_\varepsilon^{2p-2\varepsilon} + 2\beta u_\varepsilon^{p-\varepsilon} v_\varepsilon^{p-\varepsilon} + \mu_2 v_\varepsilon^{2p-2\varepsilon}) < 0,
\end{aligned}
$$

we get that $\gamma = 0$ and so $I'_\varepsilon(u_\varepsilon, v_\varepsilon) = 0$. By Lemma 5.7, we see that $u_\varepsilon \not\equiv 0$ and $v_\varepsilon \not\equiv 0$. This means that $(u_\varepsilon, v_\varepsilon)$ is a ground state solution of (5.47). Recall that $u_\varepsilon, v_\varepsilon \geq 0$ are radially symmetric nonincreasing. By regularity theory and the maximum principle, we see that $u_\varepsilon, v_\varepsilon > 0$ in $B(0, 1)$, $u_\varepsilon, v_\varepsilon \in C^2(B(0, 1))$ and are radially symmetric decreasing. □

We are now in the position to conclude the proof of Theorem 5.6.

Proof (Completion of the proof of Theorem 5.6 (2)) Recalling (5.48), for any $(u, v) \in \mathscr{M}'(1)$, there exists $t_\varepsilon > 0$ such that $(t_\varepsilon u, t_\varepsilon v) \in \mathscr{M}'_\varepsilon$ with $t_\varepsilon \to 1$ as $\varepsilon \to 0$. Then

$$
\limsup_{\varepsilon\to0} B_\varepsilon \leq \limsup_{\varepsilon\to0} I_\varepsilon(t_\varepsilon u, t_\varepsilon v) = I(u, v), \quad \forall (u, v) \in \mathscr{M}'(1).
$$

By Lemma 5.6, we have

$$
\limsup_{\varepsilon\to0} B_\varepsilon \leq B'(1) = B'. \tag{5.53}
$$

By Theorem 5.7, we may let $(u_\varepsilon, v_\varepsilon)$ be a positive ground state solution of (5.49), which is radially symmetric decreasing. By $I'_\varepsilon(u_\varepsilon, v_\varepsilon)(u_\varepsilon, v_\varepsilon) = 0$ and Sobolev inequality, it is easily seen that

$$\frac{2p - 2\varepsilon}{p - \varepsilon - 1} B_\varepsilon = \int_{B(0,1)} (|\nabla u_\varepsilon|^2 + |\nabla v_\varepsilon|^2) \geq C_0, \quad \forall 0 < \varepsilon \leq \frac{p-1}{2}, \qquad (5.54)$$

where C_0 is a positive constant independent of ε. Consequently, $u_\varepsilon, v_\varepsilon$ are uniformly bounded in $H_0^1(B(0, 1))$. Passing to a subsequence, we may assume that $u_\varepsilon \rightharpoonup u_0$ and $v_\varepsilon \rightharpoonup v_0$ weakly in $H_0^1(B(0, 1))$. Then (u_0, v_0) is a solution of

$$\begin{cases} -\Delta u = \mu_1 |u|^{2p-2}u + \beta |u|^{p-2}u|v|^p, & x \in B(0, 1), \\ -\Delta v = \mu_2 |v|^{2p-2}v + \beta |v|^{p-2}v|u|^p, & x \in B(0, 1), \\ u, v \in H_0^1(B(0, 1)). \end{cases} \qquad (5.55)$$

Assume by contradiction that $\|u_\varepsilon\|_\infty + \|v_\varepsilon\|_\infty$ is uniformly bounded, then by the dominated convergent theorem, we obtain

$$\lim_{\varepsilon \to 0} \int_{B(0,1)} u_\varepsilon^{2p-2\varepsilon} = \int_{B(0,1)} u_0^{2p}, \quad \lim_{\varepsilon \to 0} \int_{B(0,1)} v_\varepsilon^{2p-2\varepsilon} = \int_{B(0,1)} v_0^{2p},$$

$$\lim_{\varepsilon \to 0} \int_{B(0,1)} u_\varepsilon^{p-\varepsilon} v_\varepsilon^{p-\varepsilon} = \int_{B(0,1)} u_0^p v_0^p.$$

Combining these with $I'_\varepsilon(u_\varepsilon, v_\varepsilon) = I'(u_0, v_0) = 0$, it is standard to show that $u_\varepsilon \to u_0$ and $v_\varepsilon \to v_0$ strongly in $H_0^1(B(0, 1))$. Then by (5.54), we see that $(u_0, v_0) \neq (0, 0)$. Moreover, $u_0 \geq 0, v_0 \geq 0$. We may assume that $u_0 \neq 0$. Then the strong maximum principle gives $u_0 > 0$ in $B(0, 1)$. Combining these with the Pohozaev identity, we easily conclude

$$0 < \int_{\partial B(0,1)} (|\nabla u_0|^2 + |\nabla v_0|^2)(x \cdot v)\, d\sigma = 0,$$

a contradiction. Here, v denotes the outward unit normal vector on $\partial B(0, 1)$. Therefore, $\|u_\varepsilon\|_\infty + \|v_\varepsilon\|_\infty \to \infty$ as $\varepsilon \to 0$. We will use a blow-up analysis. Noting that $u_\varepsilon(0) = \max_{B(0,1)} u_\varepsilon(x)$ and $v_\varepsilon(0) = \max_{B(0,1)} v_\varepsilon(x)$, we define $K_\varepsilon := \max\{u_\varepsilon(0), v_\varepsilon(0)\}$, then $K_\varepsilon \to +\infty$. Define

$$U_\varepsilon(x) = K_\varepsilon^{-1} u_\varepsilon(K_\varepsilon^{-\alpha_\varepsilon} x), \quad V_\varepsilon(x) = K_\varepsilon^{-1} v_\varepsilon(K_\varepsilon^{-\alpha_\varepsilon} x), \quad \alpha_\varepsilon = p - 1 - \varepsilon.$$

Then

$$1 = \max\{U_\varepsilon(0), V_\varepsilon(0)\} = \max\left\{\max_{x \in B(0, K_\varepsilon^{\alpha_\varepsilon})} U_\varepsilon(x), \max_{x \in B(0, K_\varepsilon^{\alpha_\varepsilon})} V_\varepsilon(x)\right\} \quad (5.56)$$

and U_ε, V_ε satisfy

$$\begin{cases} -\Delta U_\varepsilon = \mu_1 U_\varepsilon^{2p-2\varepsilon-1} + \beta U_\varepsilon^{p-1-\varepsilon} V_\varepsilon^{p-\varepsilon}, & x \in B(0, K_\varepsilon^{\alpha_\varepsilon}), \\ -\Delta V_\varepsilon = \mu_2 V_\varepsilon^{2p-2\varepsilon-1} + \beta V_\varepsilon^{p-1-\varepsilon} U_\varepsilon^{p-\varepsilon}, & x \in B(0, K_\varepsilon^{\alpha_\varepsilon}). \end{cases}$$

Since

$$\int_{\mathbb{R}^N} |\nabla U_\varepsilon|^2 \, dx = K_\varepsilon^{-(N-2)\varepsilon} \int_{\mathbb{R}^N} |\nabla u_\varepsilon|^2 \, dx \le \int_{\mathbb{R}^N} |\nabla u_\varepsilon|^2 \, dx,$$

we see that $\{(U_\varepsilon, V_\varepsilon)\}_{n \ge 1}$ is bounded in $D^{1,2}(\mathbb{R}^N) \times D^{1,2}(\mathbb{R}^N) = D$. By elliptic estimates and up to a subsequence, we may assume that $(U_\varepsilon, V_\varepsilon) \to (U, V) \in D$ uniformly in $C_{\text{loc}}^2(\mathbb{R}^N, \mathbb{R}^2)$ as $\varepsilon \to 0$, and (U, V) satisfies (5.14), namely $I'(U, V) = 0$. Moreover, $U \ge 0$, $V \ge 0$ are radially symmetric nonincreasing. By (5.56), we have $(U, V) \ne (0, 0)$, and so $(U, V) \in \mathcal{M}'$. Then, we deduce from (5.53) that

$$B' \le I(U, V) = \left(\frac{1}{2} - \frac{1}{2p}\right) \int_{\mathbb{R}^N} (|\nabla U|^2 + |\nabla V|^2) \, dx$$

$$\le \liminf_{\varepsilon \to 0} \left(\frac{1}{2} - \frac{1}{2p - 2\varepsilon}\right) \int_{B(0, K_\varepsilon^{\alpha_\varepsilon})} (|\nabla U_\varepsilon|^2 + |\nabla V_\varepsilon|^2) \, dx$$

$$\le \liminf_{\varepsilon \to 0} \left(\frac{1}{2} - \frac{1}{2p - 2\varepsilon}\right) \int_{B(0,1)} (|\nabla u_\varepsilon|^2 + |\nabla v_\varepsilon|^2) \, dx$$

$$= \liminf_{\varepsilon \to 0} B_\varepsilon \le B'.$$

This implies that $I(U, V) = B'$. By (5.51), we have that $U \not\equiv 0$ and $V \not\equiv 0$. By the strong maximum principle, $U > 0$ and $V > 0$ are radially symmetric decreasing. We also have $(U, V) \in \mathcal{M}$, and so $I(U, V) \ge B \ge B'$, namely

$$I(U, V) = B = B', \quad (5.57)$$

and (U, V) is a positive ground state solution of (5.14), which is radially symmetric decreasing.

Finally, we show the existence of $(k(\beta), l(\beta))$ for $\beta > 0$ small. Recalling (5.20)–(5.21), We denote $\alpha_i(k, l)$ by $\alpha_i(k, l, \beta)$ here. Define $k(0) = \mu_1^{-\frac{1}{p-1}}$ and $l(0) = \mu_2^{-\frac{1}{p-1}}$, then $\alpha_i(k(0), l(0), 0) = 0$ for $i = 1, 2$. A direct computation gives

$$\partial_k \alpha_1(k(0), l(0), 0) = (p-1)\mu_1 k(0)^{p-2} > 0,$$

$$\partial_l \alpha_2(k(0), l(0), 0) = (p-1)\mu_2 l(0)^{p-2} > 0,$$

$$\partial_l \alpha_1(k(0), l(0), 0) = \partial_k \alpha_2(k(0), l(0), 0) = 0,$$

which implies that

$$\det \begin{pmatrix} \partial_k \alpha_1(k(0), l(0), 0) & \partial_l \alpha_1(k(0), l(0), 0) \\ \partial_k \alpha_2(k(0), l(0), 0) & \partial_l \alpha_2(k(0), l(0), 0) \end{pmatrix} > 0.$$

Therefore, by the implicit function theorem, $k(\beta), l(\beta)$ are well defined and class C^1 on $(-\beta_2, \beta_2)$ for some $\beta_2 > 0$, and $\alpha_i(k(\beta), l(\beta), \beta) \equiv 0, i = 1, 2$. This implies that $(\sqrt{k(\beta)}U_{\varepsilon,y}, \sqrt{l(\beta)}U_{\varepsilon,y})$ is a positive solution of (5.14). Moreover,

$$\lim_{\beta \to 0} \left(k(\beta) + l(\beta)\right) = k(0) + l(0) = \mu_1^{\frac{2-N}{2}} + \mu_2^{\frac{2-N}{2}},$$

that is, there exists $0 < \beta_1 \leq \beta_2$, such that

$$k(\beta) + l(\beta) > \min \left\{ \mu_1^{\frac{2-N}{2}}, \mu_2^{\frac{2-N}{2}} \right\}, \quad \forall \beta \in (0, \beta_1).$$

Combining this with (5.42) and (5.51), we have

$$I(U, V) = B' = B < I(\sqrt{k(\beta)}U_{\varepsilon,y}, \sqrt{l(\beta)}U_{\varepsilon,y}), \quad \forall \beta \in (0, \beta_1),$$

that is, $(\sqrt{k(\beta)}U_{\varepsilon,y}, \sqrt{l(\beta)}U_{\varepsilon,y})$ is different positive solution of (5.14) with respect to (U, V). This completes the proof. $\qquad \square$

Finally, we conclude this section by studying the following properties of (U, V) obtained in Theorem 5.6.

Proposition 5.1 *Assume that $\beta > 0$. Let (U, V) be a positive radially symmetric ground state solution of (5.14) obtained in Theorem 5.6. Then there exists $C > 0$ such that*

$$U(x) + V(x) \leq C(1 + |x|)^{2-N}, \quad |\nabla U(x)| + |\nabla V(x)| \leq C(1 + |x|)^{1-N}.$$

Proof Define the Kelvin transformation:

$$U^*(x) := |x|^{2-N} U\left(\frac{x}{|x|^2}\right), \quad V^*(x) := |x|^{2-N} V\left(\frac{x}{|x|^2}\right).$$

Then $U^*, V^* \in D^{1,2}(\mathbb{R}^N)$ and (U^*, V^*) satisfies the same system (5.14). By a standard Brezis–Kato type argument [17], we see that $U^*, V^* \in L^\infty(\mathbb{R}^N)$. Therefore, there exists $C > 0$ such that

$$U(x) + V(x) \le C|x|^{2-N}. \tag{5.58}$$

On the other hand, we note that U, V are radially symmetric decreasing. We also have $U, V \in L^\infty(\mathbb{R}^N)$, and so

$$U(x) + V(x) \le C(1 + |x|)^{2-N}.$$

Moreover, the standard elliptic regularity theory gives $U, V \in C^2(\mathbb{R}^N)$. We write $U(|x|) = U(x)$ for convenience. Then

$$(r^{N-1}U_r)_r = -r^{N-1}(\mu_1 U^{2^*-1} + \beta U^{2^*/2-1} V^{2^*/2}),$$

and so for any $R \ge 1$, we derive from (5.58) that

$$R^{N-1}|U_r(R)| \le |U_r(1)| + \int_1^R r^{N-1}(\mu_1 U^{2^*-1} + \beta U^{2^*/2-1} V^{2^*/2})\, dr$$

$$\le C + C \int_1^{+\infty} r^{N-1} r^{-N-2}\, dr \le C.$$

Therefore, it is easy to see that $|\nabla U(x)| \le C(1+|x|)^{1-N}$ for some $C > 0$. Similarly, $|\nabla V(x)| \le C(1+|x|)^{1-N}$. $\qquad\square$

5.3 The General Case $\lambda_1 \ne \lambda_2$

In this section, we give the proof of Theorem 5.3. Without loss of generality, we assume that $-\lambda_1(\Omega) < \lambda_1 \le \lambda_2 < 0$. Recalling the definition of A in (5.5), we have $A > 0$ as before. As pointed out before, by Brezis and Nirenberg [19] the Brezis–Nirenberg problem (5.3) has a positive least energy solution $u_{\mu_i} \in C^2(\Omega) \cap C(\overline{\Omega})$ with energy

$$\frac{1}{N}\left(\frac{\lambda_1(\Omega) + \lambda_i}{\lambda_1(\Omega)}\right)^{\frac{N}{2}} \mu_i^{\frac{2-N}{2}} S^{\frac{N}{2}} \le B_{\mu_i} := \frac{1}{2}\int_\Omega (|\nabla u_{\mu_i}|^2 + \lambda_i u_{\mu_i}^2) - \frac{\mu_i}{2^*}\int_\Omega u_{\mu_i}^{2^*}$$

$$< \frac{1}{N}\mu_i^{\frac{2-N}{2}} S^{N/2}, \quad i = 1, 2. \tag{5.59}$$

The following lemma is the counterpart of Lemma 4.2.

Lemma 5.8 *Let $\beta < 0$, then*

$$A < \min\left\{ B_{\mu_1} + \frac{1}{N}\mu_2^{\frac{2-N}{2}} S^{N/2}, B_{\mu_2} + \frac{1}{N}\mu_1^{\frac{2-N}{2}} S^{N/2}, B \right\}.$$

Proof The idea of this proof is similar to Lemma 4.2, but some arguments are more delicate. Let $\beta < 0$ and take $t_0 > 0$ such that

$$\frac{N}{2} B_{\mu_1} t^2 - \frac{N}{4p} B_{\mu_1} t^{2p} + \frac{1}{N}\left(\frac{\mu_1}{2}\right)^{\frac{2-N}{2}} S^{\frac{N}{2}} < 0, \quad \forall\, t > t_0. \tag{5.60}$$

Since $u_{\mu_1} \in C(\overline{\Omega})$ and $u_{\mu_1} \equiv 0$ on $\partial\Omega$, there exists $B(y_0, 2R) = \{x \mid |x - y_0| \leq 2R\} \subset \Omega$ such that

$$\delta := \max_{B(y_0, 2R)} u_{\mu_1} \leq \min\left\{ \left(\frac{\mu_2}{2|\beta|}\right)^{\frac{1}{p-1}}, \left(\frac{\lambda_1 + \lambda_1(\Omega)}{2|\beta|}\right)^{\frac{1}{p-1}} \right\}. \tag{5.61}$$

Let $\psi \in C_0^1(B(y_0, 2R))$ be a function with $0 \leq \psi \leq 1$ and $\psi \equiv 1$ for $|x - y_0| \leq R$. Define $v_\varepsilon = \psi U_{\varepsilon, y_0}$, where U_{ε, y_0} is defined in (5.16) and (5.17). Then by Brezis and Nirenberg [19] or Willem [90, Lemma 1.46], we have the following inequalities

$$\int_\Omega |\nabla v_\varepsilon|^2 = S^{\frac{N}{2}} + O(\varepsilon^{N-2}), \quad \int_\Omega |v_\varepsilon|^{2^*} = S^{\frac{N}{2}} + O(\varepsilon^N), \tag{5.62}$$

$$\int_\Omega |v_\varepsilon|^2 \geq C\varepsilon^2 + O(\varepsilon^{N-2}). \tag{5.63}$$

Moreover, since $N \geq 5$, we have

$$\int_\Omega v_\varepsilon^{\frac{N}{N-2}} \, dx \leq \int_{B(y_0, 2R)} U_{\varepsilon, y_0}^{\frac{N}{N-2}} \, dx = C \int_{B(0, 2R)} \left(\frac{\varepsilon}{\varepsilon^2 + |x|^2}\right)^{N/2} dx$$

$$\leq C\varepsilon^{N/2}\left(\ln\frac{2R}{\varepsilon} + 1\right) = o(\varepsilon^2). \tag{5.64}$$

Since $\text{supp}(v_\varepsilon) \subset B(y_0, 2R)$, by (5.61) we have for $t, s > 0$ that

$$2|\beta| t^p s^p \int_\Omega u_{\mu_1}^p v_\varepsilon^p \leq 2|\beta| \delta^{p-1} t^p s^p \int_\Omega u_{\mu_1} v_\varepsilon^p$$

$$\leq |\beta| \delta^{p-1} t^{2p} \int_\Omega u_{\mu_1}^2 + |\beta| \delta^{p-1} s^{2p} \int_\Omega v_\varepsilon^{2p}$$

$$\leq \frac{|\beta| \delta^{p-1}}{\lambda_1 + \lambda_1(\Omega)} t^{2p} \int_\Omega (|\nabla u_{\mu_1}|^2 + \lambda_1 u_{\mu_1}^2) + |\beta| \delta^{p-1} s^{2p} \int_\Omega v_\varepsilon^{2p}$$

$$\leq \frac{1}{2} t^{2p} \int_\Omega \mu_1 u_{\mu_1}^{2p} + \frac{1}{2} s^{2p} \int_\Omega \mu_2 v_\varepsilon^{2p}, \qquad (5.65)$$

and so

$$E(tu_{\mu_1}, sv_\varepsilon) = \frac{1}{2} t^2 \int_\Omega (|\nabla u_{\mu_1}|^2 + \lambda_1 u_{\mu_1}^2) + \frac{1}{2} s^2 \int_\Omega (|\nabla v_\varepsilon|^2 + \lambda_2 v_\varepsilon^2)$$

$$- \frac{1}{2p} \int_\Omega (t^{2p} \mu_1 u_{\mu_1}^{2p} + 2t^p s^p \beta u_{\mu_1}^p v_\varepsilon^p + s^{2p} \mu_2 v_\varepsilon^{2p})$$

$$\leq \frac{1}{2} t^2 \int_\Omega (|\nabla u_{\mu_1}|^2 + \lambda_1 u_{\mu_1}^2) - \frac{1}{4p} t^{2p} \int_\Omega \mu_1 u_{\mu_1}^{2p}$$

$$+ \frac{1}{2} s^2 \int_\Omega (|\nabla v_\varepsilon|^2 + \lambda_2 v_\varepsilon^2) - \frac{\mu_2}{4p} s^{2p} \int_\Omega v_\varepsilon^{2p}$$

$$= f(t) + g(s). \qquad (5.66)$$

By (5.62)–(5.63), it is standard to check that (cf. [19, 90])

$$\max_{s>0} g(s) < \frac{1}{N} \left(\frac{\mu_2}{2}\right)^{\frac{2-N}{2}} S^{\frac{N}{2}} \quad \text{for } \varepsilon \text{ small enough.} \qquad (5.67)$$

On the other hand, (5.59) gives

$$f(t) = \frac{N}{2} B_{\mu_1} t^2 - \frac{N}{4p} B_{\mu_1} t^{2p}.$$

Combining these with (5.60), we get that

$$f(t) + g(s) < 0, \quad \forall t > t_0, \, s > 0,$$

and so it follows from (5.66) that

$$\max_{t,s>0} E(tu_{\mu_1}, sv_\varepsilon) = \max_{0<t\leq t_0, s>0} E(tu_{\mu_1}, sv_\varepsilon).$$

Define

$$g_\varepsilon(s) := \frac{1}{2} s^2 \int_\Omega (|\nabla v_\varepsilon|^2 + \lambda_2 v_\varepsilon^2) \, dx - \frac{s^{2p}}{2p} \int_\Omega \mu_2 v_\varepsilon^{2p} \, dx, \quad s > 0.$$

Obviously, there exists a unique $s(\varepsilon) > 0$ such that $g'_\varepsilon(s(\varepsilon)) = 0$ with

$$s(\varepsilon)^{2p-2} = \frac{\int_\Omega (|\nabla v_\varepsilon|^2 + \lambda_2 v_\varepsilon^2)\, dx}{\int_\Omega \mu_2 v_\varepsilon^{2p}\, dx} \geq \left(1 + \frac{\lambda_2}{\lambda_1(\Omega)}\right) \frac{\int_\Omega |\nabla v_\varepsilon|^2\, dx}{\mu_2 \int_{\mathbb{R}^N} U_{\varepsilon,y_0}^{2p}\, dx}$$

$$= \left(1 + \frac{\lambda_2}{\lambda_1(\Omega)}\right) \frac{S^{N/2} + O(\varepsilon^{N-2})}{\mu_2 S^{N/2}}$$

$$\geq \frac{1}{2\mu_2}\left(1 + \frac{\lambda_2}{\lambda_1(\Omega)}\right) =: s_0^{2p-2}, \quad \text{for } \varepsilon \text{ small enough.}$$

Therefore, since g_ε is increasing for $0 < s \leq s(\varepsilon)$, we have $g_\varepsilon(s) < g_\varepsilon(s_0)$ and so $E(tu_{\mu_1}, sv_\varepsilon) < E(tu_{\mu_1}, s_0 v_\varepsilon)$ for any $0 < s < s_0$. In particular,

$$\max_{t,s>0} E(tu_{\mu_1}, sv_\varepsilon) = \max_{0<t\leq t_0, s\geq s_0} E(tu_{\mu_1}, sv_\varepsilon). \tag{5.68}$$

For $0 < t \leq t_0$ and $s \geq s_0$, we derive from (5.61), (5.64), and $p < 2$ that

$$|\beta| t^p s^p \int_\Omega u_{\mu_1}^p v_\varepsilon^p \leq |\beta| t_0^p \delta^p s_0^{p-2} s^2 \int_\Omega v_\varepsilon^p \leq Cs^2 \cdot o(\varepsilon^2),$$

Thus,

$$E(tu_{\mu_1}, sv_\varepsilon) = \frac{1}{2}t^2 \int_\Omega (|\nabla u_{\mu_1}|^2 + \lambda_1 u_{\mu_1}^2) + \frac{1}{2}s^2 \int_\Omega (|\nabla v_\varepsilon|^2 + \lambda_2 v_\varepsilon^2)$$

$$- \frac{1}{2p}\int_\Omega (t^{2p}\mu_1 u_{\mu_1}^{2p} + 2t^p s^p \beta u_{\mu_1}^p v_\varepsilon^p + s^{2p}\mu_2 v_\varepsilon^{2p})$$

$$\leq \frac{1}{2}t^2 \int_\Omega (|\nabla u_{\mu_1}|^2 + \lambda_1 u_{\mu_1}^2) - \frac{1}{2p}t^{2p}\int_\Omega \mu_1 u_{\mu_1}^{2p}$$

$$+ \frac{1}{2}s^2 \left(\int_\Omega (|\nabla v_\varepsilon|^2 + \lambda_2 v_\varepsilon^2) + o(\varepsilon^2)\right) - \frac{1}{2p}s^{2p}\mu_2 \int_\Omega v_\varepsilon^{2p}$$

$$=: f_1(t) + g_1(s). \tag{5.69}$$

We note that $\max_{t>0} f_1(t) = f_1(1) = B_{\mu_1}$. Besides, by (5.62)–(5.63) and $\lambda_2 < 0$, it is easy to prove that

$$\max_{s>0} g_1(s) < \frac{1}{N}\mu_2^{\frac{2-N}{2}} S^{N/2} \quad \text{for } \varepsilon \text{ small enough.}$$

Combining these with (5.68) and (5.69), we conclude

$$\max_{t,s>0} E(tu_{\mu_1}, sv_\varepsilon) = \max_{0<t\leq t_0, s\geq s_0} E(tu_{\mu_1}, sv_\varepsilon)$$

$$\leq \max_{t>0} f_1(t) + \max_{s>0} g_1(s)$$

$$< B_{\mu_1} + \frac{1}{N}\mu_2^{\frac{2-N}{2}} S^{N/2} \quad \text{for } \varepsilon \text{ small enough.} \quad (5.70)$$

Now, we claim that there exists $t_\varepsilon, s_\varepsilon > 0$ such that $(t_\varepsilon u_{\mu_1}, s_\varepsilon v_\varepsilon) \in \mathcal{N}$. Similarly as (5.65), we have

$$\left(\int_\Omega \beta u_{\mu_1}^p v_\varepsilon^p \, dx\right)^2 \leq |\beta|^2 \delta^{2p-2} \left(\int_\Omega u_{\mu_1} v_\varepsilon \, dx\right)^2$$

$$\leq |\beta|^2 \delta^{2p-2} \int_\Omega u_{\mu_1}^2 \, dx \int_\Omega v_\varepsilon^{2p} \, dx$$

$$\leq \frac{|\beta|^2 \delta^{2p-2}}{(\lambda_1(\Omega) + \lambda_1)\mu_2} \int_\Omega \mu_1 u_{\mu_1}^{2p} \, dx \int_\Omega \mu_2 v_\varepsilon^{2p} \, dx$$

$$< \int_\Omega \mu_1 u_{\mu_1}^{2p} \, dx \int_\Omega \mu_2 v_\varepsilon^{2p} \, dx.$$

For convenience, we denote

$$D_1 = \int_\Omega \mu_1 u_{\mu_1}^{2p} \, dx, \quad D_2 = \int_\Omega \beta u_{\mu_1}^p v_\varepsilon^p \, dx,$$

$$D_3 = \int_\Omega \mu_2 v_\varepsilon^{2p} \, dx, \quad D_4 = \int_\Omega (|\nabla v_\varepsilon|^2 + \lambda_2 v_\varepsilon^2) \, dx.$$

Then $D_2 < 0$ and $D_1 D_3 - D_2^2 > 0$. Furthermore, $(tu_{\mu_1}, sv_\varepsilon) \in \mathcal{N}$ for some $t, s > 0$ is equivalent to

$$t^{2-p}D_1 = t^p D_1 + s^p D_2, \quad s^{2-p}D_4 = s^p D_3 + t^p D_2, \quad s, t > 0. \quad (5.71)$$

Recall that $1 < p = \frac{N}{N-2} < 2$. By $s^p = (t^{2-p} - t^p)D_1/D_2 > 0$, we have $t > 1$. Therefore, (5.71) is equivalent to

$$f_3(t) := D_4 \left(\frac{D_1}{|D_2|}(1 - t^{2-2p})\right)^{\frac{2-p}{p}} - \frac{D_1 D_3 - D_2^2}{|D_2|}t^{2p-2} + \frac{D_1 D_3}{|D_2|} = 0, \quad t > 1. \quad (5.72)$$

Since $f_3(1) > 0$ and $\lim\limits_{t \to +\infty} f_3(t) < 0$, it follows that (5.72) has a solution $t > 1$. Hence, (5.71) has a solution $t_\varepsilon > 0, s_\varepsilon > 0$. That is, $(t_\varepsilon u_{\mu_1}, s_\varepsilon v_\varepsilon) \in \mathcal{N}$ and from (5.70) we get

$$A \leq E(t_\varepsilon u_{\mu_1}, s_\varepsilon v_\varepsilon) \leq \max_{t,s>0} E(t u_{\mu_1}, s v_\varepsilon) < B_{\mu_1} + \frac{1}{N} \mu_2^{\frac{2-N}{2}} S^{N/2}.$$

A similar argument also shows $A < B_{\mu_2} + \frac{1}{N} \mu_1^{\frac{2-N}{2}} S^{N/2}$. Finally, by (5.41) and (5.59), we have

$$B > \max \left\{ B_{\mu_1} + \frac{1}{N} \mu_2^{\frac{2-N}{2}} S^{N/2}, B_{\mu_2} + \frac{1}{N} \mu_1^{\frac{2-N}{2}} S^{N/2} \right\},$$

which completes the proof. $\qquad\square$

Lemma 5.9 *Assume that $\beta < 0$, then there exists $C_2 > C_1 > 0$, such that for any $(u, v) \in \mathcal{N}$ with $E(u, v) \leq B$, there holds*

$$C_1 \leq \int_\Omega |u|^{2p}\, dx, \int_\Omega |v|^{2p}\, dx \leq C_2.$$

Proof This follows directly from

$$\frac{\lambda_1(\Omega) + \lambda_1}{\lambda_1(\Omega)} S \left(\int_\Omega |u|^{2p} \right)^{\frac{1}{p}} \leq \int_\Omega (|\nabla u|^2 + \lambda_1 u^2) \leq \mu_1 \int_\Omega |u|^{2p},$$

$$\frac{\lambda_1(\Omega) + \lambda_2}{\lambda_1(\Omega)} S \left(\int_\Omega |v|^{2p} \right)^{\frac{1}{p}} \leq \int_\Omega (|\nabla v|^2 + \lambda_2 v^2) \leq \mu_1 \int_\Omega |v|^{2p},$$

$E(u, v) \leq B$ and (5.5). $\qquad\square$

The following lemma is the counterpart of Brezis–Lieb Lemma [90] for (u, v).

Lemma 5.10 *Let $u_n \rightharpoonup u, v_n \rightharpoonup v$ in $H_0^1(\Omega)$ as $n \to \infty$, then passing to a subsequence, there holds*

$$\lim_{n \to \infty} \int_\Omega \left(|u_n|^p |v_n|^p - |u_n - u|^p |v_n - v|^p - |u|^p |v|^p \right) dx = 0.$$

Proof Noting $2p = 2^*$, we have

$$u_n \to u, v_n \to v \quad \text{strongly in } L^q(\Omega), \quad \forall 0 < q < 2p,$$
$$u_n \rightharpoonup u, v_n \rightharpoonup v \quad \text{weakly in } L^{2p}(\Omega).$$

Fix any $t \in [0, 1]$. First, we claim that

$$|u_n - tu|^{p-2}(u_n - tu)|v_n|^p \rightharpoonup (1-t)^{p-1}|u|^{p-2}u|v|^p \text{ weakly in } L^{\frac{2p}{2p-1}}(\Omega). \tag{5.73}$$

Since the map $h: L^{q_1}(\Omega) \to L^{q_1/q_2}(\Omega)$ with $h(s) = |s|^{q_2-1}s$ is continuous, so

$$|u_n - tu|^{p-2}(u_n - tu) \to (1-t)^{p-1}|u|^{p-2}u \text{ strongly in } L^q(\Omega), \quad \forall 0 < q < \frac{2p}{p-1},$$

$$|v_n|^p \to |v|^p \text{ strongly in } L^q(\Omega), \quad \forall 0 < q < 2.$$

Then for any $1 \leq q < \frac{2p}{2p-1}$, one has

$$|u_n - tu|^{p-2}(u_n - tu)|v_n|^p \to (1-t)^{p-1}|u|^{p-2}u|v|^p \text{ strongly in } L^q(\Omega).$$

Since $|u_n - tu|^{p-2}(u_n - tu)|v_n|^p$ is uniformly bounded in $L^{\frac{2p}{2p-1}}(\Omega)$, passing to a subsequence, we may assume that $|u_n - tu|^{p-2}(u_n - tu)|v_n|^p \rightharpoonup w$ weakly in $L^{\frac{2p}{2p-1}}(\Omega)$. Then for any $\varphi \in C_0^\infty(\mathbb{R}^N)$, we have

$$\int_\Omega w\varphi = \lim_{n \to \infty} \int_\Omega |u_n - tu|^{p-2}(u_n - tu)|v_n|^p\varphi = \int_\Omega (1-t)^{p-1}|u|^{p-2}u|v|^p\varphi,$$

which implies $w = (1-t)^{p-1}|u|^{p-2}u|v|^p$, namely (5.73) holds. Similarly, we can show that $|u_n - u|^p|v_n - tv|^{p-2}(v_n - tv) \rightharpoonup 0$ weakly in $L^{\frac{2p}{2p-1}}(\Omega)$. Therefore, by (5.73), the Fubini theorem and the dominated convergent theorem,

$$\int_\Omega \left(|u_n|^p|v_n|^p - |u_n - u|^p|v_n - v|^p\right) dx$$

$$= p \int_\Omega \int_0^1 |u_n - tu|^{p-2}(u_n - tu)|v_n|^p u \, dt \, dx$$

$$+ p \int_\Omega \int_0^1 |u_n - u|^p|v_n - tv|^{p-2}(v_n - tv)v \, dt \, dx$$

$$= p \int_0^1 \int_\Omega |u_n - tu|^{p-2}(u_n - tu)|v_n|^p u \, dx \, dt$$

$$+ p \int\limits_0^1 \int\limits_\Omega |u_n - u|^p |v_n - tv|^{p-2}(v_n - tv)v \, dx \, dt$$

$$\to p \int\limits_0^1 \int\limits_\Omega (1-t)^{p-1}|u|^p |v|^p \, dx \, dt = \int_\Omega |u|^p |v|^p \, dx, \quad \text{as } n \to \infty.$$

This completes the proof.   □

Clearly, people might prove Lemma 5.10 via different techniques. We refer the reader to [58] for a different proof, which is essentially the same as that of Brezis–Lieb Lemma [90]. Now we can begin the proof of Theorem 5.3.

Proof (Proof of Theorem 5.3 for $\beta < 0$) The main idea of the proof is similar to the proof of Theorem 4.3 in case $N = 4$, but as we will see, some different ideas are needed. Assume that $\beta < 0$. By the Ekeland variational principle (cf. [81]), there exists a minimizing sequence $\{(u_n, v_n)\} \subset \mathcal{N}$ satisfying

$$E(u_n, v_n) \leq \min \left\{A + \tfrac{1}{n}, B\right\}, \tag{5.74}$$

$$E(u, v) \geq E(u_n, v_n) - \tfrac{1}{n}\|(u_n, v_n) - (u, v)\|, \quad \forall(u, v) \in \mathcal{N}. \tag{5.75}$$

Here, $\|(u, v)\| := (\int_\Omega (|\nabla u|^2 + |\nabla v|^2) \, dx)^{1/2}$ is the norm of H. Then $\{(u_n, v_n)\}$ is bounded in H. For any $(\varphi, \phi) \in H$ with $\|\varphi\|, \|\phi\| \leq 1$ and each $n \in \mathbb{N}$, we define the functions h_n and $g_n: \mathbb{R}^3 \to \mathbb{R}$ by

$$h_n(t, s, l) = \int\limits_\Omega |\nabla(u_n + t\varphi + su_n)|^2 + \lambda_1 \int\limits_\Omega |u_n + t\varphi + su_n|^2$$

$$- \mu_1 \int\limits_\Omega |u_n + t\varphi + su_n|^{2p} - \beta \int\limits_\Omega |u_n + t\varphi + su_n|^p |v_n + t\phi + lv_n|^p,$$

$$\tag{5.76}$$

$$g_n(t, s, l) = \int\limits_\Omega |\nabla(v_n + t\phi + lv_n)|^2 + \lambda_2 \int\limits_\Omega |v_n + t\phi + lv_n|^2$$

$$- \mu_2 \int\limits_\Omega |v_n + t\phi + lv_n|^{2p} - \beta \int\limits_\Omega |u_n + t\varphi + su_n|^p |v_n + t\phi + lv_n|^p.$$

$$\tag{5.77}$$

Denote $\mathbf{0} = (0, 0, 0)$. Clearly, $h_n, g_n \in C^1(\mathbb{R}^3, \mathbb{R})$ and $h_n(\mathbf{0}) = g_n(\mathbf{0}) = 0$. Moreover,

$$\frac{\partial h_n}{\partial s}(0) = -(2p-2)\mu_1 \int_\Omega |u_n|^{2p} - (p-2)\beta \int_\Omega |u_n|^p |v_n|^p,$$

$$\frac{\partial h_n}{\partial l}(0) = \frac{\partial g_n}{\partial s}(0) = -p\beta \int_\Omega |u_n|^p |v_n|^p \, dx,$$

$$\frac{\partial g_n}{\partial l}(0) = -(2p-2)\mu_2 \int_\Omega |v_n|^{2p} - (p-2)\beta \int_\Omega |u_n|^p |v_n|^p.$$

Define a matrix

$$F_n := \begin{pmatrix} \frac{\partial h_n}{\partial s}(0) & \frac{\partial h_n}{\partial l}(0) \\ \frac{\partial g_n}{\partial s}(0) & \frac{\partial g_n}{\partial l}(0) \end{pmatrix}.$$

Since $\beta < 0$ and $(u_n, v_n) \in \mathcal{N}$, we have

$$\mu_1 \int_\Omega |u_n|^{2p} = \int_\Omega (|\nabla u_n|^2 + \lambda_1 u_n^2) + |\beta| \int_\Omega |u_n|^p |v_n|^p,$$

$$\mu_2 \int_\Omega |v_n|^{2p} = \int_\Omega (|\nabla v_n|^2 + \lambda_2 v_n^2) + |\beta| \int_\Omega |u_n|^p |v_n|^p,$$

Then, we deduce from Lemma 5.9 that

$$\det(F_n) \geq (2p-2)^2 \int_\Omega (|\nabla u_n|^2 + \lambda_1 |u_n|^2) \int_\Omega (|\nabla v_n|^2 + \lambda_2 |v_n|^2)$$

$$\geq CS^2 \left(\int_\Omega |u_n|^{2p} \right)^{\frac{1}{p}} \left(\int_\Omega |v_n|^{2p} \right)^{\frac{1}{p}} \geq C > 0, \tag{5.78}$$

where C is independent of n. Then by repeating the progress of Step 2 of Lemma 2.6 in Chap. 2, we can prove that

$$\lim_{n \to +\infty} E'(u_n, v_n) = 0. \tag{5.79}$$

Since $\{(u_n, v_n)\}$ is bounded in H, we may assume that $(u_n, v_n) \rightharpoonup (u, v)$ weakly in H. Passing to a subsequence, we may assume that

$u_n \rightharpoonup u, v_n \rightharpoonup v$, weakly in $L^{2p}(\Omega)$,

$|u_n|^{q-1} u_n \rightharpoonup |u|^{q-1} u, |v_n|^{q-1} v_n \rightharpoonup |v|^{q-1} v$, weakly in $L^{2p/q}(\Omega)$, $1 < q < 2p$,

$u_n \to u, v_n \to v$, strongly in $L^2(\Omega)$.

Thus, by (5.79) we have $E'(u, v) = 0$. Set $\omega_n = u_n - u$ and $\sigma_n = v_n - v$. Then by Brezis–Lieb Lemma (cf. [18, 90]), there holds

$$|u_n|_{2p}^{2p} = |u|_{2p}^{2p} + |\omega_n|_{2p}^{2p} + o(1), \quad |v_n|_{2p}^{2p} = |v|_{2p}^{2p} + |\sigma_n|_{2p}^{2p} + o(1). \quad (5.80)$$

Notice that $(u_n, v_n) \in \mathcal{N}$ and $E'(u, v) = 0$. Combining these with (5.80) and Lemma 5.10, we get that

$$\int_\Omega |\nabla \omega_n|^2 - \int_\Omega (\mu_1 |\omega_n|^{2p} + \beta |\omega_n|^p |\sigma_n|^p) = o(1), \quad (5.81)$$

$$\int_\Omega |\nabla \sigma_n|^2 - \int_\Omega (\mu_2 |\sigma_n|^{2p} + \beta |\omega_n|^p |\sigma_n|^p) = o(1), \quad (5.82)$$

$$E(u_n, v_n) = E(u, v) + I(\omega_n, \sigma_n) + o(1). \quad (5.83)$$

Passing to a subsequence, we may assume that

$$\lim_{n \to +\infty} \int_\Omega |\nabla \omega_n|^2 = b_1, \quad \lim_{n \to +\infty} \int_\Omega |\nabla \sigma_n|^2 = b_2.$$

Then by (5.81) and (5.82), we have $I(\omega_n, \sigma_n) = \frac{1}{N}(b_1 + b_2) + o(1)$. Letting $n \to +\infty$ in (5.83), we get that

$$0 \leq E(u, v) \leq E(u, v) + \frac{1}{N}(b_1 + b_2) = \lim_{n \to +\infty} E(u_n, v_n) = A. \quad (5.84)$$

Case 1. $u \equiv 0, v \equiv 0$.

By Lemma 5.9, (5.80), and (5.84), we have $0 < b_1 < +\infty$ and $0 < b_2 < +\infty$, and we may assume that both $\omega_n \not\equiv 0$ and $\sigma_n \not\equiv 0$ for n large. Moreover, (5.81) and (5.82) give

$$\int_\Omega \mu_1 |\omega_n|^{2p} \int_\Omega \mu_2 |\sigma_n|^{2p} - \left(\beta \int_\Omega |\omega_n|^p |\sigma_n|^p \right)^2 > 0, \quad \text{for } n \text{ large.}$$

Then by a similar argument as Lemma 5.8, there exists $t_n, s_n > 0$ for n large such that $(t_n \omega_n, s_n \sigma_n) \in \mathcal{M}$. Up to a subsequence, we claim that

$$\lim_{n \to +\infty} (|t_n - 1| + |s_n - 1|) = 0. \quad (5.85)$$

This conclusion is obvious in case $N = 4$ and $p = 2$ (see Chap. 4), but it is not trivial in our general case $N \geq 5$ here. Denote

$$B_{n,1} = \int_\Omega |\nabla \omega_n|^2 \to b_1, \quad B_{n,2} = \int_\Omega |\nabla \sigma_n|^2 \to b_2,$$

$$C_{n,1} = \int_\Omega \mu_1 |\omega_n|^{2p}, \quad C_{n,2} = \int_\Omega \mu_2 |\sigma_n|^{2p},$$

$$D_n = |\beta| \int_\Omega |\omega_n|^p |\sigma_n|^p.$$

Passing to a subsequence, we may assume that $C_{n,1} \to c_1 < +\infty, C_{n,2} \to c_2 < +\infty$ and $D_n \to d < +\infty$. By (5.81)–(5.82), we have

$$c_1 = b_1 + d \geq b_1 > 0, \quad c_2 = b_2 + d \geq b_2 > 0, \tag{5.86}$$

$$t_n^2 B_{n,1} = t_n^{2p} C_{n,1} - t_n^p s_n^p D_n, \quad s_n^2 B_{n,2} = s_n^{2p} C_{n,2} - t_n^p s_n^p D_n. \tag{5.87}$$

This implies that

$$t_n^{2p-2} \geq \frac{B_{n,1}}{C_{n,1}} \to \frac{b_1}{c_1} > 0, \quad s_n^{2p-2} \geq \frac{B_{n,2}}{C_{n,2}} \to \frac{b_2}{c_2} > 0. \tag{5.88}$$

Assume that, up to a subsequence, $t_n \to +\infty$ as $n \to \infty$, then by

$$t_n^{2p} C_{n,1} - t_n^2 B_{n,1} = s_n^{2p} C_{n,2} - s_n^2 B_{n,2},$$

we also have $s_n \to +\infty$. Consequently,

$$\begin{aligned}
d^2 = \lim_{n\to\infty} D_n^2 &= \lim_{n\to\infty} \frac{t_n^p C_{n,1} - t_n^{2-p} B_{n,1}}{s_n^p} \cdot \frac{s_n^p C_{n,2} - s_n^{2-p} B_{n,2}}{t_n^p} \\
&= \lim_{n\to\infty} (C_{n,1} - t_n^{2-2p} B_{n,1})(C_{n,2} - s_n^{2-2p} B_{n,2}) \\
&= c_1 c_2 = (b_1 + d)(b_2 + d) > d^2,
\end{aligned}$$

a contradiction. Therefore, t_n, s_n are uniformly bounded. Up to a subsequence, by (5.88) we may assume that $t_n \to t_\infty \geq (b_1/c_1)^{\frac{1}{2p-2}} > 0$ and $s_n \to s_\infty \geq (b_2/c_2)^{\frac{1}{2p-2}} > 0$. It follows from (5.87) that

$$s_\infty^p d = t_\infty^p c_1 - t_\infty^{2-p} b_1, \quad t_\infty^p d = s_\infty^p c_2 - s_\infty^{2-p} b_2.$$

If $d = 0$, then $c_i = b_i$, and so $t_\infty = s_\infty = 1$, namely (5.85) holds. Now we consider the case $d > 0$. Define $f(t) = t^p c_1 - t^{2-p} b_1$, then for $t \geq (b_1/c_1)^{\frac{1}{2p-2}}$, we have

$$f'(t) = pc_1 t^{p-1} - (2-p)b_1 t^{1-p} > (2-p)t^{1-p}(c_1 t^{2p-2} - b_1) \geq 0,$$

that is, f is increasing with respect to $t \geq (b_1/c_1)^{\frac{1}{2p-2}}$. If $t_\infty < 1$, then

$$s_\infty^p d = f(t_\infty) < f(1) = c_1 - b_1 = d,$$

namely $s_\infty < 1$, and we derive from (5.86) that

$$
\begin{aligned}
d^2 &= \frac{t_\infty^p c_1 - t_\infty^{2-p} b_1}{s_\infty^p} \cdot \frac{s_\infty^p c_2 - s_\infty^{2-p} b_2}{t_\infty^p} = (c_1 - t_\infty^{2-2p} b_1)(c_2 - s_\infty^{2-2p} b_2) \\
&= (d + b_1 - t_\infty^{2-2p} b_1)(d + b_2 - s_\infty^{2-2p} b_2) < d^2,
\end{aligned}
$$

a contradiction. If $t_\infty > 1$, since $1 \geq (b_1/c_1)^{\frac{1}{2p-2}}$, we have

$$s_\infty^p d = f(t_\infty) > f(1) = c_1 - b_1 = d,$$

namely $s_\infty > 1$, and so

$$d^2 = (d + b_1 - t_\infty^{2-2p} b_1)(d + b_2 - s_\infty^{2-2p} b_2) > d^2,$$

also a contradiction. Therefore, $t_\infty = s_\infty = 1$ and (5.85) holds. This implies that

$$\frac{1}{N}(b_1 + b_2) = \lim_{n \to +\infty} I(\omega_n, \sigma_n) = \lim_{n \to +\infty} I(t_n \omega_n, s_n \sigma_n) \geq B.$$

Combining this with (5.84), we conclude $A \geq B$, which is a contradiction with Lemma 5.8. Therefore, Case 1 is impossible.

Case 2. Either $u \not\equiv 0$, $v \equiv 0$ or $u \equiv 0$, $v \not\equiv 0$.

Without loss of generality, we assume that $u \not\equiv 0$ and $v \equiv 0$. Then $b_2 > 0$. By Case 1, we may assume that $b_1 = 0$. Then $\lim_{n \to +\infty} \int_\Omega |\omega_n|^p |\sigma_n|^p = 0$, and so

$$\int_\Omega |\nabla \sigma_n|^2 = \int_\Omega \mu_2 |\sigma_n|^{2p} + o(1) \leq \mu_2 S^{-p} \left(\int_\Omega |\nabla \sigma_n|^2 \right)^p + o(1).$$

This implies that $b_2 \geq \mu_2^{\frac{2-N}{2}} S^{N/2}$. Meanwhile, clearly u is a nontrivial solution of $-\Delta u + \lambda_1 u = \mu_1 |u|^{2^*-2} u$ in Ω, so (5.59) yields $E(u, 0) \geq B_{\mu_1}$. By (5.84), we get

$$A \geq B_{\mu_1} + \frac{1}{N} b_2 \geq B_{\mu_1} + \frac{1}{N} \mu_2^{\frac{2-N}{2}} S^{N/2},$$

a contradiction with Lemma 5.8. Therefore, Case 2 is impossible.

Since Cases 1 and 2 are both impossible, we have $u \not\equiv 0$ and $v \not\equiv 0$, namely $(u, v) \in \mathcal{N}$. By (5.84), we have $E(u, v) = A$. Consequently, $(|u|, |v|) \in \mathcal{M}$ and $E(|u|, |v|) = A$. Lemma 5.5 indicates that $(|u|, |v|)$ is a solution of (5.1). The maximum principle yields $|u|, |v| > 0$ in Ω. Therefore, $(|u|, |v|)$ is a positive ground state solution of (5.1). This completes the proof. $\qquad\square$

It remains to prove Theorem 5.3 for the case $\beta > 0$. Let $\beta > 0$ and define the mountain pass minimax value

$$\mathcal{A} := \inf_{h \in \Gamma} \max_{t \in [0,1]} E(h(t)), \tag{5.89}$$

where $\Gamma = \{h \in C([0, 1], H): h(0) = (0, 0), E(h(1)) < 0\}$. By (5.4), we see that for any $(u, v) \in H \setminus \{(0, 0)\}$,

$$\max_{t > 0} E(tu, tv) = E(t_{u,v}u, t_{u,v}v)$$

$$= \frac{1}{N} t_{u,v}^2 \int_{\Omega} (|\nabla u|^2 + \lambda_1 u^2 + |\nabla v|^2 + \lambda_2 v^2)$$

$$= \frac{1}{N} t_{u,v}^{2^*} \int_{\Omega} (\mu_1 |u|^{2p} + 2\beta |u|^p |v|^p + \mu_2 |v|^{2p}), \tag{5.90}$$

where $t_{u,v} > 0$ satisfies

$$t_{u,v}^{2p-2} = \frac{\int_{\Omega} (|\nabla u|^2 + \lambda_1 u^2 + |\nabla v|^2 + \lambda_2 v^2)}{\int_{\Omega} (\mu_1 |u|^{2p} + 2\beta |u|^p |v|^p + \mu_2 |v|^{2p})}. \tag{5.91}$$

Note that $(t_{u,v}u, t_{u,v}v) \in \mathcal{N}'$, where

$$\mathcal{N}' := \left\{ (u, v) \in H \setminus \{(0, 0)\} \,\middle|\, G(u, v) := \int_{\Omega} (|\nabla u|^2 + \lambda_1 u^2 + |\nabla v|^2 + \lambda_2 v^2) \right.$$

$$\left. - \int_{\Omega} (\mu_1 |u|^{2p} + 2\beta |u|^p |v|^p + \mu_2 |v|^{2p}) = 0 \right\}, \tag{5.92}$$

it is easy to check that

$$\mathcal{A} = \inf_{H \ni (u,v) \neq (0,0)} \max_{t > 0} E(tu, tv) = \inf_{(u,v) \in \mathcal{N}'} E(u, v). \tag{5.93}$$

Noting $\mathcal{N} \subset \mathcal{N}'$, one has that $\mathcal{A} \leq A$. Similarly as (5.39), we have $\mathcal{A} > 0$.

Lemma 5.11 Let $\beta > 0$, then $\mathcal{A} < \min\{B_{\mu_1}, B_{\mu_2}, B\}$.

Proof Step 1. We prove that $\mathscr{A} < B$.

Without loss of generality, we may assume that $0 \in \Omega$. Then there exists $\rho > 0$ such that $B(0, 2\rho) = \{x \mid |x| \leq 2\rho\} \subset \Omega$. Let $\psi \in C_0^1(B(0, 2\rho))$ be a nonnegative function with $0 \leq \psi \leq 1$ and $\psi \equiv 1$ for $|x| \leq \rho$. Recalling that (U, V) in Theorem 5.6, we define

$$(U_\varepsilon(x), V_\varepsilon(x)) := \left(\varepsilon^{-\frac{N-2}{2}} U\left(\frac{x}{\varepsilon}\right), \ \varepsilon^{-\frac{N-2}{2}} V\left(\frac{x}{\varepsilon}\right) \right).$$

Then, it is easy to see that

$$\int_{\mathbb{R}^N} |\nabla U_\varepsilon|^2 = \int_{\mathbb{R}^N} |\nabla U|^2, \quad \int_{\mathbb{R}^N} |U_\varepsilon|^{2^*} = \int_{\mathbb{R}^N} |U|^{2^*},$$

$$\int_{\mathbb{R}^N} |\nabla V_\varepsilon|^2 = \int_{\mathbb{R}^N} |\nabla V|^2, \quad \int_{\mathbb{R}^N} |V_\varepsilon|^{2^*} = \int_{\mathbb{R}^N} |V|^{2^*}.$$

Define

$$(u_\varepsilon, v_\varepsilon) := (\psi U_\varepsilon, \psi V_\varepsilon). \tag{5.94}$$

First, we claim the following inequalities

$$\int_\Omega |\nabla u_\varepsilon|^2 \leq \int_{\mathbb{R}^N} |\nabla U|^2 + O(\varepsilon^{N-2}), \tag{5.95}$$

$$\int_\Omega |u_\varepsilon|^{2^*} \geq \int_{\mathbb{R}^N} |U|^{2^*} + O(\varepsilon^N), \tag{5.96}$$

$$\int_\Omega |u_\varepsilon|^{\frac{2^*}{2}} |v_\varepsilon|^{\frac{2^*}{2}} \geq \int_{\mathbb{R}^N} |U|^{\frac{2^*}{2}} |V|^{\frac{2^*}{2}} + O(\varepsilon^N), \tag{5.97}$$

$$\int_\Omega |u_\varepsilon|^2 \geq C\varepsilon^2 + O(\varepsilon^{N-2}), \tag{5.98}$$

where C is a positive constant.

Let $0 < \varepsilon \ll \rho$. By Proposition 5.1, we have

$$\int_\Omega |\nabla \psi|^2 |U_\varepsilon|^2 \, dx \leq C \int_{\rho \leq |x| \leq 2\rho} \varepsilon^{2-N} U^2(x/\varepsilon) \, dx$$

$$\leq C\varepsilon^2 \int_{\rho/\varepsilon \leq |x| \leq 2\rho/\varepsilon} U^2(x) \, dx$$

$$\leq C\varepsilon^2 \int\limits_{\rho/\varepsilon \leq |x| \leq 2\rho/\varepsilon} |x|^{4-2N} \, dx = O(\varepsilon^{N-2});$$

$$\int\limits_{\Omega} |\nabla U_\varepsilon|^2 |\psi|^2 \, dx \leq \int\limits_{\mathbb{R}^N} |\nabla U_\varepsilon|^2 = \int\limits_{\mathbb{R}^N} |\nabla U|^2;$$

$$\left| \int\limits_{\Omega} \psi U_\varepsilon \nabla \psi \nabla U_\varepsilon \right| \leq C \int\limits_{\rho \leq |x| \leq 2\rho} |\nabla U_\varepsilon| |U_\varepsilon| \, dx$$

$$\leq C \int\limits_{\rho \leq |x| \leq 2\rho} \varepsilon^{1-N} |\nabla_x U(x/\varepsilon)| |U(x/\varepsilon)| \, dx$$

$$= C\varepsilon \int\limits_{\rho/\varepsilon \leq |x| \leq 2\rho/\varepsilon} |\nabla U(x)| |U(x)| \, dx$$

$$\leq C\varepsilon \int\limits_{\rho/\varepsilon \leq |x| \leq 2\rho/\varepsilon} |x|^{3-2N} \, dx = O(\varepsilon^{N-2}).$$

Therefore,

$$\int\limits_{\Omega} |\nabla u_\varepsilon|^2 \, dx = \int\limits_{\mathbb{R}^N} |\nabla U_\varepsilon|^2 |\psi|^2 + \int\limits_{\Omega} |\nabla \psi|^2 |U_\varepsilon|^2 + 2 \int\limits_{\Omega} \psi U_\varepsilon \nabla \psi \nabla U_\varepsilon$$

$$\leq \int\limits_{\mathbb{R}^N} |\nabla U|^2 \, dx + O(\varepsilon^{N-2}),$$

namely (5.95) holds. Note that

$$\int\limits_{\mathbb{R}^N} (1 - \psi^{2^*}) |U_\varepsilon|^{2^*} \, dx \leq \int\limits_{|x| \geq \rho} \varepsilon^{-N} |U(x/\varepsilon)|^{2^*} \, dx = \int\limits_{|x| \geq \rho/\varepsilon} |U(x)|^{2^*} \, dx$$

$$\leq C \int\limits_{|x| \geq \rho/\varepsilon} |x|^{-2N} \, dx = O(\varepsilon^N),$$

then

$$\int\limits_{\Omega} |u_\varepsilon|^{2^*} \, dx = \int\limits_{\mathbb{R}^N} |U_\varepsilon|^{2^*} \, dx - \int\limits_{\mathbb{R}^N} (1 - \psi^{2^*}) |U_\varepsilon|^{2^*} \, dx$$

$$\geq \int\limits_{\mathbb{R}^N} |U|^{2^*} \, dx + O(\varepsilon^N),$$

namely (5.96) holds. Similarly, (5.97) holds. Note that

$$\int_{\Omega} |u_{\varepsilon}|^2 \, dx \geq \int_{|x| \leq \rho} \varepsilon^{2-N} |U(x/\varepsilon)|^2 \, dx$$

$$= \varepsilon^2 \int_{\mathbb{R}^N} U^2 \, dx - \varepsilon^2 \int_{|x| \geq \rho/\varepsilon} U^2(x) \, dx$$

$$\geq C\varepsilon^2 - C\varepsilon^2 \int_{|x| \geq \rho/\varepsilon} |x|^{4-2N} \, dx = C\varepsilon^2 + O(\varepsilon^{N-2}),$$

namely (5.98) holds. Similarly, we have

$$\int_{\Omega} |\nabla v_{\varepsilon}|^2 \leq \int_{\mathbb{R}^N} |\nabla V|^2 + O(\varepsilon^{N-2}), \tag{5.99}$$

$$\int_{\Omega} |v_{\varepsilon}|^{2^*} \geq \int_{\mathbb{R}^N} |V|^{2^*} + O(\varepsilon^N), \tag{5.100}$$

$$\int_{\Omega} |v_{\varepsilon}|^2 \geq C\varepsilon^2 + O(\varepsilon^{N-2}). \tag{5.101}$$

Recalling that $I(U, V) = B$, we have

$$NB = \int_{\mathbb{R}^N} \left(|\nabla U|^2 + |\nabla V|^2 \right) = \int_{\mathbb{R}^N} \mu_1 U^{2^*} + 2\beta U^{\frac{2^*}{2}} V^{\frac{2^*}{2}} + \mu_2 V^{2^*}.$$

Combining this with (5.95)–(5.101) and recalling that $\lambda_1, \lambda_2 < 0$, $2p = 2^*$, $N \geq 5$, we have for any $t > 0$ that

$$E(tu_{\varepsilon}, tv_{\varepsilon}) = \frac{1}{2} t^2 \int_{\Omega} \left(|\nabla u_{\varepsilon}|^2 + \lambda_1 u_{\varepsilon}^2 + |\nabla v_{\varepsilon}|^2 + \lambda_2 v_{\varepsilon}^2 \right)$$

$$- \frac{1}{2p} t^{2p} \int_{\Omega} \left(\mu_1 u_{\varepsilon}^{2p} + 2\beta u_{\varepsilon}^p v_{\varepsilon}^p + \mu_2 v_{\varepsilon}^{2p} \right)$$

$$\leq \frac{1}{2} \left(\int_{\mathbb{R}^N} \left(|\nabla U|^2 + |\nabla V|^2 \right) - C\varepsilon^2 + O(\varepsilon^{N-2}) \right) t^2$$

$$- \frac{1}{2^*} \left(\int_{\mathbb{R}^N} \left(\mu_1 U^{2^*} + 2\beta U^{\frac{2^*}{2}} V^{\frac{2^*}{2}} + \mu_2 V^{2^*} \right) + O(\varepsilon^N) \right) t^{2^*}$$

$$= \frac{1}{2} \left(NB - C\varepsilon^2 + O(\varepsilon^{N-2}) \right) t^2 - \frac{1}{2^*} \left(NB + O(\varepsilon^N) \right) t^{2^*}$$

$$\leq \frac{1}{N}\left(NB - C\varepsilon^2 + O(\varepsilon^{N-2})\right)\left(\frac{NB - C\varepsilon^2 + O(\varepsilon^{N-2})}{NB + O(\varepsilon^N)}\right)^{\frac{N-2}{2}}$$

$$< B \quad \text{for } \varepsilon > 0 \text{ small enough.} \tag{5.102}$$

Hence, for $\varepsilon > 0$ small enough, there holds

$$\mathscr{A} \leq \max_{t>0} E(tu_\varepsilon, tv_\varepsilon) < B. \tag{5.103}$$

Step 2. we shall prove $\mathscr{A} < B_{\mu_1}$.

This proof is similar to Lemma 5.7. Recalling (5.59) and (5.91), we define $t(s) := t_{u_{\mu_1}, su_{\mu_1}}$, that is,

$$t(s)^{2p-2} = \frac{\int_\Omega(|\nabla u_{\mu_1}|^2 + \lambda_1 u_{\mu_1}^2 + s^2|\nabla u_{\mu_1}|^2 + s^2\lambda_2 u_{\mu_1}^2)}{(\mu_1 + 2\beta|s|^p + \mu_2|s|^{2p})\int_\Omega u_{\mu_1}^{2p}}.$$

Note that $t(0) = 1$ and $1 < p < 2$. A direct computation gives

$$\lim_{s\to 0} \frac{t'(s)}{|s|^{p-2}s} = -\frac{2p\beta}{(2p-2)\mu_1},$$

that is,

$$t'(s) = -\frac{2p\beta}{(2p-2)\mu_1}|s|^{p-2}s(1 + o(1)), \quad \text{as } s \to 0,$$

and so

$$t(s) = 1 - \frac{2\beta}{(2p-2)\mu_1}|s|^p(1 + o(1)), \quad \text{as } s \to 0.$$

This implies that

$$t(s)^{2p} = 1 - \frac{2p\beta}{(p-1)\mu_1}|s|^p(1 + o(1)), \quad \text{as } s \to 0.$$

Therefore, we deduce from (5.90), $\frac{2p}{2p-2} = N/2$ and $B_{\mu_1} = \frac{1}{N}\int_\Omega u_{\mu_1}^{2p}$ that

$$\mathscr{A} \leq E\left(t(s)u_{\mu_1}, t(s)su_{\mu_1}\right) = \frac{t(s)^{2p}}{N}(\mu_1 + 2\beta|s|^p + \mu_2|s|^{2p})\int_\Omega u_{\mu_1}^{2p}$$

$$= B_{\mu_1} - 2\beta\left(\frac{1}{2} - \frac{1}{N}\right)|s|^p\int_\Omega |u_{\mu_1}|^{2p} + o(|s|^p)$$

$$< B_{\mu_1} \quad \text{as } |s| > 0 \text{ small enough,}$$

namely $\mathscr{A} < B_{\mu_1}$. A similar argument also shows $\mathscr{A} < B_{\mu_2}$. This completes the proof. $\square$

Now we can finish the proof of Theorem 5.3.

Proof (Proof of Theorem 5.3 for the case $\beta > 0$) Assume that $\beta > 0$. By the classical mountain pass theorem, there exists $\{(u_n, v_n)\} \subset H$ such that

$$\lim_{n \to +\infty} E(u_n, v_n) = \mathscr{A}, \qquad \lim_{n \to +\infty} E'(u_n, v_n) = 0.$$

As before, we may assume that $(u_n, v_n) \rightharpoonup (u, v)$ weakly in H. Setting $\omega_n = u_n - u$ and $\sigma_n = v_n - v$ and using the same symbols as in the proof of Theorem 5.3 for the case $\beta < 0$, we see that $E'(u, v) = 0$ and (5.81)–(5.83) also hold. Moreover,

$$0 \leq E(u, v) \leq E(u, v) + \frac{1}{N}(b_1 + b_2) = \lim_{n \to +\infty} E(u_n, v_n) = \mathscr{A}. \qquad (5.104)$$

Case 1. $u \equiv 0, v \equiv 0$.

By (5.104), we have $b_1 + b_2 > 0$. Then we may assume that $(\omega_n, \sigma_n) \neq (0, 0)$ for n large. Recalling $\mathscr{M}'$ in (5.45), by (5.81)–(5.82), it is easy to check that there exists $t_n > 0$ such that $(t_n \omega_n, t_n \sigma_n) \in \mathscr{M}'$ and $t_n \to 1$ as $n \to \infty$. Consequently, by (5.57) and (5.104) we have

$$\mathscr{A} = \frac{1}{N}(b_1 + b_2) = \lim_{n \to +\infty} I(\omega_n, \sigma_n) = \lim_{n \to +\infty} I(t_n \omega_n, t_n \sigma_n) \geq B' = B,$$

a contradiction with Lemma 5.11. Therefore, Case 1 is impossible.

Case 2. Either $u \not\equiv 0, v \equiv 0$ or $u \equiv 0, v \not\equiv 0$.

Without loss of generality, we may assume that $u \not\equiv 0$ and $v \equiv 0$. Then, u is a nontrivial solution of $-\Delta u + \lambda_1 u = \mu_1 |u|^{2^*-2} u$ in Ω, and so $\mathscr{A} \geq E(u, 0) \geq B_{\mu_1}$, a contradiction with Lemma 5.11. Therefore, Case 2 is also impossible.

Since Cases 1 and 2 are both impossible, by repeating the progress of proving Theorem 4.3 for $\beta > \beta_2$ in Chap. 4, we conclude that $(|u|, |v|)$ is a positive ground state solution of (5.1). $\square$

5.4 Uniqueness of Ground State Solutions

In this section, we turn back to the symmetric case $-\lambda_1(\Omega) < \lambda_1 = \lambda_2 = \lambda < 0$ and prove Theorem 5.2. For this, we assume $\beta \geq (p - 1) \max\{\mu_1, \mu_2\}$. Define $g \colon [(p - 1) \max\{\mu_1, \mu_2\}, +\infty)$ by

$$g(\beta) := (p - 1)\mu_1 \mu_2 \beta^{2/p-2} + \beta^{2/p}. \qquad (5.105)$$

Then

$$g'(\beta) = \frac{2}{p}\beta^{2/p-3}\left(\beta^2 - (p-1)^2\mu_1\mu_2\right) > 0, \quad \forall \beta > (p-1)\max\{\mu_1, \mu_2\}.$$

A direct computation gives

$$g\left((p-1)\max\{\mu_1, \mu_2\}\right) \leq p(p-1)^{\frac{2}{p}-1}\max\left\{\mu_1^{2/p}, \mu_2^{2/p}\right\}.$$

Therefore, there exists a unique $\beta_0 \geq (p-1)\max\{\mu_1, \mu_2\}$ such that

$$g(\beta_0) = p(p-1)^{\frac{2}{p}-1}\max\left\{\mu_1^{2/p}, \mu_2^{2/p}\right\}, \quad \text{and} \tag{5.106}$$

$$g(\beta) > p(p-1)^{\frac{2}{p}-1}\max\left\{\mu_1^{2/p}, \mu_2^{2/p}\right\}, \quad \forall \beta > \beta_0. \tag{5.107}$$

Moreover,

$$\beta_0 = (p-1)\max\{\mu_1, \mu_2\}, \quad \text{if } \mu_1 = \mu_2. \tag{5.108}$$

Lemma 5.12 *Assume that $\beta > \beta_0$, where β_0 is defined in (5.106). Let (k_0, l_0) be in Lemma 5.1. Then $p\mu_1 k_0^{p-1} < 1$ and $p\mu_2 l_0^{p-1} < 1$.*

Proof Let $k_1 = (p\mu_1)^{\frac{1}{p-1}}$, then (5.22) gives

$$l_1 := h_1(k_1) = \left[\frac{p-1}{p\beta(p\mu_1)^{\frac{2-p}{2(p-1)}}}\right]^{2/p}.$$

By (5.107) and direct computations, we get that

$$\begin{aligned}
\alpha_2(k_1, l_1) &= \mu_2 l_1^{p-1} + \beta k_1^{p/2} l_1^{p/2-1} - 1 \\
&= \frac{1}{l_1}\left[\mu_2 l_1^p + k_1(1 - \mu_1 k_1^{p-1})\right] - 1 = \frac{1}{l_1}\left[\mu_2 l_1^p + \frac{p-1}{p}k_1\right] - 1 \\
&= \left[\frac{p\beta(p\mu_1)^{\frac{2-p}{2(p-1)}}}{p-1}\right]^{\frac{2}{p}}\left\{\mu_2\left[\frac{p-1}{p\beta(p\mu_1)^{\frac{2-p}{2(p-1)}}}\right]^2 + \frac{p-1}{p}(p\mu_1)^{-\frac{1}{p-1}}\right\} - 1 \\
&= (p-1)^{1-2/p}p^{-1}\mu_1^{-2/p}g(\beta) - 1 > 0.
\end{aligned}$$

Combining this with Lemma 5.3 we have $k_1 > k_0$, namely $p\mu_1 k_0^{p-1} < 1$. Similarly, let $l_2 = (p\mu_2)^{\frac{1}{p-1}}$, then

$$\alpha_1(h_2(l_2), l_2) = (p-1)^{1-2/p}p^{-1}\mu_2^{-2/p}g(\beta) - 1 > 0.$$

By Lemma 5.3 again, we have $l_2 > l_0$, and so $p\mu_2 l_0^{p-1} < 1$. $\qquad\qquad\qquad$ $\square$

Lemma 5.13 *Assume that $\beta > \beta_0$, where β_0 is defined in (5.106). Recall α_1, α_2 defined in (5.20)–(5.21), and (k_0, l_0) obtained in Lemma 5.1. Then*

$$F(k_0, l_0) := \det \begin{pmatrix} \partial_k \alpha_1(k_0, l_0) & \partial_l \alpha_1(k_0, l_0) \\ \partial_k \alpha_2(k_0, l_0) & \partial_l \alpha_2(k_0, l_0) \end{pmatrix} < 0.$$

Proof By $\alpha_1(k_0, l_0) = \alpha_2(k_0, l_0) = 0$, we have

$$\beta k_0^{p/2-2} l_0^{p/2} = k_0^{-1} - \mu_1 k_0^{p-2}, \quad \beta l_0^{p/2-2} k_0^{p/2} = l_0^{-1} - \mu_2 l_0^{p-2}.$$

Then,

$$\partial_k \alpha_1(k_0, l_0) = (p-1)\mu_1 k_0^{p-2} + (p/2 - 1)\beta k_0^{p/2-2} l_0^{p/2}$$
$$= \frac{p}{2}\mu_1 k_0^{p-2} - (1 - p/2)k_0^{-1};$$

$$\partial_l \alpha_2(k_0, l_0) = (p-1)\mu_2 l_0^{p-2} + (p/2 - 1)\beta l_0^{p/2-2} k_0^{p/2}$$
$$= \frac{p}{2}\mu_2 l_0^{p-2} - (1 - p/2)l_0^{-1};$$

$$\partial_l \alpha_1(k_0, l_0) = \partial_k \alpha_2(k_0, l_0) = \frac{p}{2}\beta k_0^{p/2-1} l_0^{p/2-1}$$
$$= \frac{p}{2}\sqrt{(k_0^{-1} - \mu_1 k_0^{p-2})(l_0^{-1} - \mu_2 l_0^{p-2})}.$$

Therefore,

$$F(k_0, l_0) = \left[\frac{p}{2}\mu_1 k_0^{p-2} - (1 - p/2)k_0^{-1}\right]\left[\frac{p}{2}\mu_2 l_0^{p-2} - (1 - p/2)l_0^{-1}\right]$$
$$- \frac{p^2}{4}\left(k_0^{-1} - \mu_1 k_0^{p-2}\right)\left(l_0^{-1} - \mu_2 l_0^{p-2}\right)$$
$$= \frac{p}{2}(p-1)k_0^{-1} l_0^{-1}\left(\mu_1 k_0^{p-1} + \mu_2 l_0^{p-1} - \frac{2}{p}\right) < 0$$

from Lemma 5.12. $\qquad\qquad\qquad\qquad\qquad\qquad\qquad\qquad\qquad\qquad$ $\square$

Lemma 5.14 *Fix any $\mu_1, \mu_2 > 0$, and $\beta > \beta_0$. Let (u_0, v_0) be a ground state solution of (5.1) with (μ_1, μ_2, β) which exists by Theorem 5.3. Recall $(\sqrt{k_0}\omega, \sqrt{l_0}\omega)$ in Theorem 5.1. Then,*

$$\int_\Omega |u_0|^{2p}\, dx = k_0^p \int_\Omega \omega^{2p}\, dx. \qquad\qquad\qquad (5.109)$$

Proof Fix any $\mu_1, \mu_2 > 0$, and $\beta > \beta_0$. We remark from (5.105)–(5.106) that $\beta_0(\mu_1, \mu_2) := \beta_0$ is completely determined by μ_1, μ_2. Hence, there exists $0 < \varepsilon < \mu_1$ such that for any $\mu \in (\mu_1 - \varepsilon, \mu_1 + \varepsilon)$, one has $\beta > \beta_0(\mu, \mu_2)$. Then by

Lemmas 5.1, 5.13, and the implicit function theorem, when μ_1 is replaced by μ, functions $k_0(\mu)$ and $l_0(\mu)$ are well defined and class C^1 for $\mu \in (\mu_1 - \varepsilon_1, \mu_1 + \varepsilon_1)$ for some $0 < \varepsilon_1 \le \varepsilon$. Recalling the definition of $E, \mathcal{N}$ and A, they all depend on μ, and we use notations $E_\mu, \mathcal{N}_\mu, A(\mu)$ in this proof when μ_1 is replaced by μ. Then $A(\mu) = (k_0(\mu) + l_0(\mu))B_1 \in C^1((\mu_1 - \varepsilon_1, \mu_1 + \varepsilon_1), \mathbb{R})$. In particular, $A'(\mu_1) := \frac{d}{d\mu}A(\mu)$ exists. Note that $A = \mathscr{A}$ by the proof of Theorem 5.3 for the case $\beta > 0$. Then by (5.93), we have

$$A(\mu) = \inf_{H \ni (u,v) \ne (0,0)} \max_{t > 0} E_\mu(tu, tv).$$

Denote

$$C = \int_\Omega (|\nabla u_0|^2 + \lambda_1 u_0^2 + |\nabla v_0|^2 + \lambda_2 v_0^2),$$

$$D = \int_\Omega (2\beta |u_0|^p |v_0|^p + \mu_2 |v_0|^{2p}), \quad G = \int_\Omega |u_0|^{2p} \, dx.$$

We note that there exists $t(\mu) > 0$ such that

$$\max_{t > 0} E_\mu(tu_0, tv_0) = E_\mu\Big(t(\mu)u_0, t(\mu)v_0\Big),$$

where $t(\mu) > 0$ satisfies $f(\mu, t(\mu)) = 0$, and

$$f(\mu, t) := t^{2p-2}(\mu G + D) - C.$$

Note that $f(\mu_1, 1) = 0$, $\frac{\partial}{\partial t}f(\mu_1, 1) = (2p - 2)(\mu_1 G + D) > 0$, and $f(\mu, t(\mu)) \equiv 0$. By the implicit function theorem, there exists $0 < \varepsilon_2 \le \varepsilon_1$, such that $t(\mu) \in C^\infty((\mu_1 - \varepsilon_2, \mu_1 + \varepsilon_2), \mathbb{R})$. By $f(\mu, t(\mu)) \equiv 0$, we easily deduce that

$$t'(\mu_1) = -\frac{G}{(2p - 2)(\mu_1 G + D)}.$$

By Taylor expansion, one has $t(\mu) = 1 + t'(\mu_1)(\mu - \mu_1) + O((\mu - \mu_1)^2)$, and so

$$t^2(\mu) = 1 + 2t'(\mu_1)(\mu - \mu_1) + O((\mu - \mu_1)^2).$$

Noting $C = \mu_1 G + D = NA(\mu_1)$, it follows from (5.90) that

$$A(\mu) \le E_\mu(t(\mu)u_0, t(\mu)v_0) = \frac{1}{N}t^2(\mu)C = t^2(\mu)A(\mu_1)$$

$$= A(\mu_1) - \frac{2GA(\mu_1)}{(2p - 2)(\mu_1 G + D)}(\mu - \mu_1) + O((\mu - \mu_1)^2)$$

$$= A(\mu_1) - \frac{G}{2p}(\mu - \mu_1) + O((\mu - \mu_1)^2),$$

Then by the same argument as Theorem 4.2 in Chap. 4, we conclude $A'(\mu_1) = -\frac{G}{2p} = -\frac{1}{2p} \int_\Omega |u_0|^{2p} \, dx$. By Theorem 5.1, $(\sqrt{k_0}\omega, \sqrt{l_0}\omega)$ is also a positive ground state solution of (5.1), so $A'(\mu_1) = -\frac{k_0^p}{2p} \int_\Omega \omega^{2p} \, dx$, namely (5.109) holds. $\square$

Now we are in a position to prove Theorem 5.2.

Proof (Proof of Theorem 5.2) Let (u, v) be any a positive ground state solution of (5.1). By Lemma 5.14, we have

$$\int_\Omega |u|^{2p} \, dx = k_0^p \int_\Omega \omega^{2p} \, dx.$$

By a similar proof of Lemma 5.14, that is, by computing $B'(\mu_2)$ and $B'(\beta)$, respectively, we can show that

$$\int_\Omega |v|^{2p} \, dx = l_0^p \int_\omega \omega^{2p} \, dx, \quad \text{and} \quad \int_\Omega |u|^p |v|^p \, dx = k_0^{p/2} l_0^{p/2} \int_\Omega \omega^{2p} \, dx.$$

Therefore,

$$\int_\Omega |u|^p |v|^p \, dx = l_0^{p/2} k_0^{-p/2} \int_\Omega |u|^{2p} \, dx, \quad \int_\Omega |u|^p |v|^p \, dx = l_0^{-p/2} k_0^{p/2} \int_\Omega |v|^{2p} \, dx.$$

In particular, by Hölder inequality, we conclude that $u = Cv$ for some constant $C > 0$. The rest proof is similar to that of Theorem 4.2, and we omit the details here. $\square$

5.5 Phase Separation and Sign-Changing Solutions of Brezis–Nirenberg Problem

In this section, we study the asymptotic behaviors of ground state solutions as $\beta \to -\infty$, and give the proofs of Theorems 5.4 and 5.5. We use notations $E_\beta, \mathcal{N}_\beta, A_\beta$ in the following. Define $B(x_0, R) := \{x \in \mathbb{R}^N : |x - x_0| < R\}$. Consider the Brezis–Nirenberg problem in a ball

$$\begin{cases} -\Delta u + \lambda_2 u = \mu_2 u^{2^*-1} \text{ in } B(0, R), \\ u > 0 \quad \text{in } B(0, R), \quad u = 0 \text{ on } \partial B(0, R), \end{cases} \tag{5.110}$$

and the corresponding functional is $J_R \colon H_0^1(B(0, R)) \to \mathbb{R}$ given by

$$J_R(u) = \frac{1}{2} \int_{B(0,R)} (|\nabla u|^2 + \lambda_2 u^2)\, dx - \frac{1}{2^*} \mu_2 \int_{B(0,R)} |u|^{2^*}\, dx. \tag{5.111}$$

We need the following sharp energy estimates, the proof of which will be given in the next section.

Theorem 5.8 (see [31]) *Let $N \geq 5$. Then there exists $R_0 > 0$ and $C_1, C_2 > 0$, such that for any $0 < R < R_0$, (5.110) has a least energy solution U_R and*

$$\frac{1}{N} \mu_2^{-\frac{N-2}{2}} S^{N/2} - C_1 R^{\frac{2N-4}{N-4}} \leq J_R(U_R) \leq \frac{1}{N} \mu_2^{-\frac{N-2}{2}} S^{N/2} - C_2 R^{\frac{2N-4}{N-4}}. \tag{5.112}$$

With the help of Theorem 5.8, we have the following lemma, which improves Lemma 5.8 in case $N \geq 6$.

Lemma 5.15 *Let $N \geq 6$. Then*

$$\sup_{\beta < 0} A_\beta < \min\left\{ B_{\mu_1} + \frac{1}{N} \mu_2^{-\frac{N-2}{2}} S^{N/2}, B_{\mu_2} + \frac{1}{N} \mu_1^{-\frac{N-2}{2}} S^{N/2} \right\}.$$

Proof Let $N \geq 6$. For any $R > 0$ small, we take $x_R \in \Omega$ with $\text{dist}(x_R, \partial\Omega) = 3R$. Then

$$|u_{\mu_1}(x)| \leq CR, \quad x \in B(x_R, 3R). \tag{5.113}$$

Let $\psi \in C_0^\infty(B(0, 2))$ with $0 \leq \psi \leq 1$ and $\psi \equiv 1$ in $B(0, 1)$. Define $\varphi_R(x) := 1 - \psi(\frac{x - x_R}{R})$, then

$$\varphi_R(x) := \begin{cases} 0 & \text{if } x \in B(x_R, R), \\ 1 & \text{if } x \in \mathbb{R}^N \setminus B(x_R, 2R), \end{cases} \quad |\nabla \varphi_R(x)| \leq C/R. \tag{5.114}$$

Define $u_R := \varphi_R u_{\mu_1}$, then by (5.113) and (5.114), it is easy to prove that

$$\int_\Omega |\nabla u_R|^2\, dx \leq \int_\Omega |\nabla u_{\mu_1}|^2\, dx + CR^N;$$

$$\int_\Omega |u_R|^2\, dx \geq \int_\Omega |u_{\mu_1}|^2\, dx - CR^{N+2};$$

$$\int_\Omega |u_R|^{2^*}\, dx \geq \int_\Omega |u_{\mu_1}|^{2^*}\, dx - CR^{N+2^*}.$$

Therefore, there exists $t_R > 0$ independent of $\beta < 0$ such that

$$
\max_{t>0} E_\beta(tu_R, 0) = E_\beta(t_R u_R, 0) = \frac{1}{N} \left(\frac{\int_\Omega(|\nabla u_R|^2 + \lambda_1 u_R^2)}{\left(\mu_1 \int_\Omega |u_R|^{2^*}\right)^{2/2^*}} \right)^{N/2}
$$

$$
\leq \frac{1}{N} \left(\frac{\int_\Omega(|\nabla u_{\mu_1}|^2 + \lambda_1 u_{\mu_1}^2) + CR^N + CR^{N+2}}{\left(\int_\Omega \mu_1 |u_{\mu_1}|^{2^*} - CR^{N+2^*}\right)^{2/2^*}} \right)^{N/2}
$$

$$
= \frac{1}{N} \left(\frac{NB_{\mu_1} + CR^N + CR^{N+2}}{\left(NB_{\mu_1} - CR^{N+2^*}\right)^{2/2^*}} \right)^{N/2}
$$

$$
\leq B_{\mu_1} + CR^N \quad \text{for } R > 0 \text{ small enough.}
$$

Recalling U_R in Theorem 5.8, we have $U_R(\cdot - x_R) \cdot u_R \equiv 0$, and so $(t_R u_R, U_R(\cdot - x_R)) \in \mathcal{N}_\beta$ for all $\beta < 0$. Since $N \geq 6$, one has $N > \frac{2N-4}{N-4}$. Consequently, we derive from Theorem 5.8 that

$$
\sup_{\beta<0} A_\beta \leq E_\beta(t_R u_R, U_R(\cdot - x_R)) = E_\beta(t_R u_R, 0) + E_\beta(0, U_R(\cdot - x_R))
$$

$$
= E_\beta(t_R u_R, 0) + J_R(U_R)
$$

$$
\leq B_{\mu_1} + CR^N + \frac{1}{N} \mu_2^{-\frac{N-2}{2}} S^{N/2} - C_2 R^{\frac{2N-4}{N-4}}
$$

$$
< B_{\mu_1} + \frac{1}{N} \mu_2^{-\frac{N-2}{2}} S^{N/2} \quad \text{for } R > 0 \text{ small enough.}
$$

A similar argument also shows $\sup_{\beta<0} A_\beta < B_{\mu_2} + \frac{1}{N} \mu_1^{-\frac{N-2}{2}} S^{N/2}$. $\square$

Now we are in a position to prove Theorem 5.4.

Proof (Proof of Theorem 5.4) This proof is similar to that of Theorem 4.4 in case $N = 4$. The novelty here is that, with the help of Lemma 5.15, we can exclude statements (1)–(2) in case $N \geq 6$. Let $\beta_n < 0, n \in \mathbb{N}$ satisfy $\beta_n \to -\infty$ as $n \to \infty$, and (u_n, v_n) be the positive ground state solutions of (5.1) with $\beta = \beta_n$. By Lemma 5.8, $E_{\beta_n}(u_n, v_n) \leq B$ and so (u_n, v_n) is uniformly bounded in H by (5.5). Passing to a subsequence, we may assume that

$$
u_n \rightharpoonup u_\infty, v_n \rightharpoonup v_\infty \text{ weakly in } H_0^1(\Omega).
$$

Then by following the proof of Theorem 4.4 in case $N = 4$, we can prove that $\int_\Omega \beta_n u_n^p v_n^p \, dx \to 0$ as $n \to \infty$, and passing to a subsequence, one of the following conclusions holds.

(1) $u_n \to u_\infty$ strongly in $H_0^1(\Omega)$ and $v_n \rightharpoonup 0$ weakly in $H_0^1(\Omega)$ (so $v_n \to 0$ for almost every $x \in \Omega$), where u_∞ is a positive least energy solution of

$$
-\Delta u + \lambda_1 u = \mu_1 |u|^{2^*-2} u, \quad u \in H_0^1(\Omega).
$$

Moreover,

$$\lim_{n\to\infty} A_{\beta_n} = B_{\mu_1} + \frac{1}{N}\mu_2^{\frac{2-N}{2}} S^{N/2}. \tag{5.115}$$

(2) $v_n \to v_\infty$ strongly in $H_0^1(\Omega)$ and $u_n \rightharpoonup 0$ weakly in $H_0^1(\Omega)$ (so $u_n \to 0$ for almost every $x \in \Omega$), where v_∞ is a positive least energy solution of

$$-\Delta v + \lambda_2 v = \mu_2 |v|^{2^*-2} v, \quad v \in H_0^1(\Omega).$$

Moreover,

$$\lim_{n\to\infty} A_{\beta_n} = B_{\mu_2} + \frac{1}{N}\mu_1^{\frac{2-N}{2}} S^{N/2}. \tag{5.116}$$

(3) $(u_n, v_n) \to (u_\infty, v_\infty)$ strongly in $H_0^1(\Omega) \times H_0^1(\Omega)$ and $u_\infty \cdot v_\infty = 0$ for almost $x \in \Omega$, where $u_\infty \not\equiv 0, v_\infty \not\equiv 0$ satisfy

$$\int_\Omega (|\nabla u_\infty|^2 + \lambda_1 u_\infty^2) = \int_\Omega \mu_1 u_\infty^{2p}, \tag{5.117}$$

$$\int_\Omega (|\nabla v_\infty|^2 + \lambda_2 v_\infty^2) = \int_\Omega \mu_2 v_\infty^{2p}, \tag{5.118}$$

$$\lim_{n\to\infty} A_{\beta_n} = E(u_\infty, v_\infty). \tag{5.119}$$

Moreover, if u_∞ and v_∞ are both continuous (we will prove this later), then $u_\infty \cdot v_\infty \equiv 0$, $u_\infty \in C(\overline{\Omega})$ is a positive least energy solution of

$$-\Delta u + \lambda_1 u = \mu_1 |u|^{2^*-2} u, \quad u \in H_0^1(\{u_\infty > 0\}),$$

and $v_\infty \in C(\overline{\Omega})$ is a positive least energy solution of

$$-\Delta v + \lambda_2 v = \mu_2 |v|^{2^*-2} v, \quad v \in H_0^1(\{v_\infty > 0\}).$$

Furthermore, both $\{v_\infty > 0\}$ and $\{u_\infty > 0\}$ are connected domains, and $\{v_\infty > 0\} = \Omega \setminus \{u_\infty > 0\}$.

Note that (5.115)–(5.116) imply that one of (1) and (2) in Theorem 5.4 does not hold in some cases. For example, if we assume that $-\lambda_1(\Omega) < \lambda_1 < \lambda_2 < 0$ and $\mu_1 = \mu_2$ in Theorem 5.4, then $B_{\mu_1} + \frac{1}{N}\mu_2^{\frac{2-N}{2}} S^{N/2} < B_{\mu_2} + \frac{1}{N}\mu_1^{\frac{2-N}{2}} S^{N/2}$, and so (2) in Theorem 5.4 does not hold, since (5.116) contradicts with Lemma 5.8.

In particular, Lemma 5.15 indicates that neither (1) nor (2) hold in case $N \geq 6$. That is, only (3) holds if $N \geq 6$. Therefore, the proof is complete by combining Lemma 5.16. $\square$

From the previous proof, it suffices to prove that both u_∞ and v_∞ are continuous and $u_\infty - v_\infty$ is a least energy sign-changing solution of (5.11). As pointed out in Remark 5.5, the following argument is completely different from that in Chap. 4 for the case $N = 4$, and also works for the case $N = 4$.

Lemma 5.16 *Let* (u_∞, v_∞) *be in conclusion* (3). *Then* $u_\infty - v_\infty$ *is a least energy sign-changing solution of* (5.11), *and* u_∞, v_∞ *are both continuous.*

Proof Consider the problem (5.11). Its related functional is

$$P(u) = \frac{1}{2} \int_\Omega (|\nabla u|^2 + \lambda_1 (u^+)^2 + \lambda_2 (u^-)^2) - \frac{1}{2^*} \int_\Omega (\mu_1 (u^+)^{2^*} + \mu_2 (u^-)^{2^*}).$$

It is standard to prove that $P \in C^1$ and its critical points are solutions of (5.11). Define

$$J_i(u) := \int_\Omega (|\nabla u|^2 + \lambda_i u^2 - \mu_i |u|^{2^*}), \quad i = 1, 2,$$

$$\mathscr{S} := \{u \in H_0^1(\Omega) : u^\pm \not\equiv 0, J_1(u^+) = 0, J_2(u^-) = 0\},$$

$$m := \inf_{u \in \mathscr{S}} P(u).$$

Then any sign-changing solutions of (5.11) belong to $\mathscr{S}$. By (5.117)–(5.118), we have $u_\infty - v_\infty \in \mathscr{S}$ and so $m \leq P(u_\infty - v_\infty) = E(u_\infty, v_\infty)$. On the other hand, for any $u \in \mathscr{S}$, we have $(u^+, u^-) \in \mathcal{N}_\beta$, for all β. Then, (5.119) yields

$$P(u_\infty - v_\infty) = E(u_\infty, v_\infty) = \lim_{n \to \infty} A_{\beta_n} \leq E(u^+, u^-) = P(u), \quad \forall u \in \mathscr{S},$$

and so $P(u_\infty - v_\infty) \leq m$. Combining these with Lemma 5.15, we obtain

$$P(u_\infty - v_\infty) = m$$

$$\left(< \min \left\{ B_{\mu_1} + \frac{1}{N} \mu_2^{\frac{2-N}{2}} S^{\frac{N}{2}}, \ B_{\mu_2} + \frac{1}{N} \mu_1^{\frac{2-N}{2}} S^{\frac{N}{2}} \right\} \quad \text{if } N \geq 6 \right). \tag{5.120}$$

Step 1. We show that $P'(u_\infty - v_\infty) = 0$, and so $u_\infty - v_\infty$ is a least energy sign-changing solution of (5.11).

Thanks to (5.120), the following argument is standard (see [22, 66] for example), and we give the details here for completeness.

Assume that $u_\infty - v_\infty$ is not a critical point of P, then there exists $\phi \in C_0^\infty(\Omega)$ such that $P'(u_\infty - v_\infty)\phi \leq -1$. Consequently, there exists $0 < \varepsilon_0 < 1/10$ such that for $|t - 1| \leq \varepsilon_0, |s - 1| \leq \varepsilon_0, |\sigma| \leq \varepsilon_0$, there holds

$$P'(tu_\infty - sv_\infty + \sigma\phi)\phi \leq -\frac{1}{2}.$$

Consider a function $0 \leq \eta \leq 1$ defined for $(t, s) \in T = [\frac{1}{2}, \frac{3}{2}] \times [\frac{1}{2}, \frac{3}{2}]$, such that

$$\eta(t, s) = 1, \quad \text{for } |t - 1| \leq \frac{\varepsilon_0}{2}, |s - 1| \leq \frac{\varepsilon_0}{2},$$
$$\eta(t, s) = 0, \quad \text{for } |t - 1| \geq \varepsilon_0 \text{ or } |s - 1| \geq \varepsilon_0.$$

Then for $|t - 1| \leq \varepsilon_0$, $|s - 1| \leq \varepsilon_0$, we have

$$P(tu_\infty - sv_\infty + \varepsilon_0 \eta(t, s)\phi)$$

$$= P(tu_\infty - sv_\infty) + \int_0^1 P'(tu_\infty - sv_\infty + \theta \varepsilon_0 \eta(t, s)\phi)[\varepsilon_0 \eta(t, s)\phi] \, d\theta$$

$$\leq P(tu_\infty - sv_\infty) - \frac{1}{2}\varepsilon_0 \eta(t, s).$$

Note that

$$\sup_{t,s>0} P(tu_\infty - sv_\infty) = P(u_\infty - v_\infty) = m,$$

and for $|t - 1| \geq \frac{\varepsilon_0}{2}$ or $|s - 1| \geq \frac{\varepsilon_0}{2}$, there exists $0 < \delta < \frac{\varepsilon_0}{2}$ such that

$$P(tu_\infty - sv_\infty) \leq m - \delta.$$

We have, for $|t - 1| \leq \frac{\varepsilon_0}{2}$, $|s - 1| \leq \frac{\varepsilon_0}{2}$, that

$$P(tu_\infty - sv_\infty + \varepsilon_0 \eta(t, s)\phi) \leq m - \frac{\varepsilon_0}{2};$$

for $\frac{\varepsilon_0}{2} \leq |t - 1| \leq \varepsilon_0$, $|s - 1| \leq \varepsilon_0$ or $\frac{\varepsilon_0}{2} \leq |s - 1| \leq \varepsilon_0$, $|t - 1| \leq \varepsilon_0$,

$$P(tu_\infty - sv_\infty + \varepsilon_0 \eta(t, s)\phi) \leq P(tu_\infty - sv_\infty) \leq m - \delta;$$

for $|t - 1| \geq \varepsilon_0$ or $|s - 1| \geq \varepsilon_0$,

$$P(tu_\infty - sv_\infty + \varepsilon_0 \eta(t, s)\phi) = P(tu_\infty - sv_\infty) \leq m - \delta.$$

So

$$\sup_{(t,s)\in T} P(tu_\infty - sv_\infty + \varepsilon_0 \eta(t, s)\phi) \leq m - \delta. \tag{5.121}$$

On the other hand, for $\varepsilon \in [0, \varepsilon_0]$, let $h_\varepsilon: T \to H_0^1(\Omega)$ by $h_\varepsilon(t, s) = tu_\infty - sv_\infty + \varepsilon \eta(t, s)\phi$, and $H_\varepsilon: T \to \mathbb{R}^2$ by

$$H_\varepsilon(t, s) = (J_1(h_\varepsilon(t, s)^+), J_2(h_\varepsilon(t, s)^-)).$$

Note that for any $(t, s) \in \partial T$, we have $\eta(t, s) = 0$ and so $h_\varepsilon(t, s) \equiv h_0(t, s) = tu_\infty - sv_\infty$. Moreover, $H_0(t, s) = (J_1(tu_\infty), J_2(sv_\infty))$. Then, it is easy to see that

$$\deg(H_{\varepsilon_0}(t, s), T, (0, 0)) = \deg(H_0(t, s), T, (0, 0)) = 1,$$

that is, there exists $(t_0, s_0) \in T$ such that $h_{\varepsilon_0}(t_0, s_0) \in \mathscr{S}$, which is a contradiction with (5.121).

Step 2. We show that u_∞ and v_∞ are continuous.

By Step 1, $u_\infty - v_\infty$ is a nontrivial solution of (5.11). Then by a Brezis–Kato argument (see [17]), we see that $u_\infty - v_\infty \in L^q(\Omega)$, $\forall q \geq 2$. In particular,

$$\mu_1 u_\infty^{2^*-1} - \mu_2 v_\infty^{2^*-1} - \lambda_1 u_\infty + \lambda_2 v_\infty \in L^q(\Omega), \quad \forall q > N.$$

Then by elliptic regularity theory, $u_\infty - v_\infty \in W^{2,q}(\Omega)$ with $q > N$. By Sobolev embedding, we have $u_\infty - v_\infty \in C(\overline{\Omega})$. Since $u_\infty = (u_\infty - v_\infty)^+$ and $v_\infty = (u_\infty - v_\infty)^-$, we see that u_∞ and v_∞ are both continuous. This completes the proof and so completes the proof of Theorem 5.4. □

Proof (Proof of Theorem 5.5) Let $N \geq 6$. In fact, by Theorem 5.4 and Lemma 5.16, we have proved that the problem (5.11) has a least energy sign-changing solution $u_\infty - v_\infty$. Obviously, Theorem 5.5 is a direct corollary by letting $\lambda_1 = \lambda_2$ and $\mu_1 = \mu_2$, and (5.13) follows directly from (5.120). □

5.6 Sharp Energy Estimates for Brezis–Nirenberg Problem

In this section, we give the proof of Theorem 5.8. In fact, we will prove a more general result. For convenience, we write the Brezis–Nirenberg problem as

$$-\Delta u = \lambda u + u^{2^*-1}, \quad u > 0 \text{ in } \Omega, \ u|_{\partial\Omega} = 0. \tag{5.122}$$

Clearly, solutions of (5.122) are critical points of a C^2 functional $I_{\lambda,\Omega} : H_0^1(\Omega) \to \mathbb{R}$, where

$$I_{\lambda,\Omega}(u) := \frac{1}{2}\int_\Omega (|\nabla u|^2 - \lambda u^2) \, dx - \frac{1}{2^*}\int_\Omega |u|^{2^*} \, dx. \tag{5.123}$$

As pointed out before, Brezis and Nirenberg [19] proved that, for $N \geq 4$ and $\lambda \in (0, \lambda_1(\Omega))$, (5.122) has a positive least energy solution u with the least energy $I_{\lambda,\Omega}(u) < \frac{1}{N}S^{N/2}$. However, there seems *no* further sharper estimates for the least energy. Here, we prove sharp energy estimates of the Brezis–Nirenberg problem in balls. Fix any $\lambda > 0$. Then $\lambda < \lambda_1(B_R)$ for small R, and so problem

$$\begin{cases} -\Delta u = \lambda u + u^{2^*-1} \text{ in } B_R, \\ u > 0 \text{ in } B_R, \quad u|_{\partial B_R} = 0, \end{cases} \tag{5.124}$$

has a positive least energy solution u_R (mountain pass solution) with energy

$$c(R) := I_{\lambda, B_R}(u_R) = \inf_{u \in H_0^1(B_R) \setminus \{0\}} \max_{t>0} I_{\lambda, B_R}(tu). \tag{5.125}$$

In a joint work with Zou [32], we proved the following result, and Theorem 5.8 is a direct corollary.

Theorem 5.9 ([32]) *Assume that $N \geq 4$ and $\lambda > 0$. Let $R_0 > 0$ such that $\lambda = \lambda_1(B_{R_0})$. Then for any $0 < R < R_0$, problem (5.124) has a least energy solution u_R. Moreover,*

(1) *if $N \geq 5$, then there exist $\widetilde{R} \in (0, R_0)$ and positive constants C_1, C_2 independent of R, such that for any $0 < R \leq \widetilde{R}$,*

$$\frac{1}{N} S^{N/2} - C_1 R^{\frac{2N-4}{N-4}} \leq I_{\lambda, B_R}(u_R) \leq \frac{1}{N} S^{N/2} - C_2 R^{\frac{2N-4}{N-4}}; \tag{5.126}$$

(2) *if $N = 4$, then there exist $\widetilde{R} \in (0, R_0)$ and positive constants C_1, C_2, C_3, C_4 independent of R, such that for any $0 < R \leq \widetilde{R}$,*

$$\frac{1}{N} S^{N/2} - C_1 e^{-C_3 R^{-2}} \leq I_{\lambda, B_R}(u_R) \leq \frac{1}{N} S^{N/2} - C_2 e^{-C_4 R^{-2}}. \tag{5.127}$$

Remark 5.8 Since 1983 when it was introduced, the Brezis–Nirenberg problem (5.122) has always be one of the focussed topics in elliptic PDEs. We believe that our Theorem 5.9 will be an important complement to the study of this classical problem. We have applied Theorem 5.9 to prove Lemma 5.15, which plays a crucial role in the proof of Theorem 5.4. We refer the reader to our paper [32] for another application of Theorem 5.9. For $N = 4$, we derive from (5.127) that, for any $k \in \mathbb{N}$,

$$\lim_{R \to 0} \frac{\frac{1}{N} S^{N/2} - I_{\lambda, B_R}(u_R)}{R^k} = 0. \tag{5.128}$$

Comparing (5.128) with (5.126), it turns out that for the Brezis–Nirenberg problem, the special case $N = 4$ is quite different from the higher dimensional case $N \geq 5$. This fact also provides an evidence that critical exponent problems might become quite different as the spatial dimensions change. This is also a motivation for us to study system (5.1) in dimensions $N = 4$ and $N \geq 5$ separately.

The proof of Theorem 5.9 seems very interesting to myself. Therefore, in the rest of this section, I would like to give the full proof of Theorem 5.9 by following the procedure in our paper [32].

First, we consider the Eq. (5.124) in the unit ball B_1, namely

$$\begin{cases} -\Delta u = \lambda u + u^{2^*-1} \text{ in } B_1, \\ u > 0 \text{ in } B_1, \quad u|_{\partial B_1} = 0. \end{cases} \tag{5.129}$$

By Brezis and Nirenberg [19], we know that (5.129) has a least energy solution U_λ for any $\lambda \in (0, \lambda_1(B_1))$. Denote $e(\lambda) := I_\lambda(U_\lambda)$, where $I_\lambda := I_{\lambda, B_1}$ is the corresponding functional of (5.129) and $I_{\lambda, \Omega}$ is defined in (5.123). Then $e(\lambda) < e(0) := \frac{1}{N} S^{N/2}$. For any $\lambda \in [0, \lambda_1(B_1))$, there holds

$$e(\lambda) = \inf_{u \in H_0^1(B_1)\setminus\{0\}} \max_{t>0} I_\lambda(tu), \tag{5.130}$$

and for any $u \in H_0^1(B_1)\setminus\{0\}$, we have

$$\max_{t>0} I_\lambda(tu) = I_\lambda(\tau_{\lambda,u} u) = \frac{1}{N} \tau_{\lambda,u}^{2^*} \int_{B_1} |u|^{2^*} \, dx, \tag{5.131}$$

where $\tau_{\lambda,u} > 0$ satisfies

$$\tau_{\lambda,u}^{2^*-2} = \frac{\int_{B_1} (|\nabla u|^2 - \lambda |u|^2) \, dx}{\int_{B_1} |u|^{2^*} \, dx}. \tag{5.132}$$

Here, the conclusion that (5.130) holds for $\lambda = 0$ is guaranteed by the fact that S is also the best constant of the embedding $H_0^1(B_1) \hookrightarrow L^{2^*}(B_1)$.

By (5.130), it is easy to prove that $e(\lambda)$ is strictly descreasing with respect to $\lambda \in [0, \lambda_1(B_1))$, and so $e'(\lambda) := \frac{de}{d\lambda}(\lambda)$ exists for almost every $\lambda \in (0, \lambda_1(B_1))$. Repeating the proof of Lemma 5.14 with minor modifications, we have

Lemma 5.17 *For any $\lambda \in [0, \lambda_1(B_1))$ such that $e'(\lambda)$ exists, there holds*

$$e'(\lambda) = -\frac{1}{2} \int_{B_1} |U_\lambda|^2 \, dx.$$

By Lemma 5.17, we can give the expression of $e(\lambda)$ in the following lemma.

Lemma 5.18 *For any $\lambda \in (0, \lambda_1(B_1))$, there holds*

$$e(\lambda) = e(0) - \frac{1}{2} \int_0^\lambda \left(\int_{B_1} |U_\mu|^2 \, dx \right) d\mu. \tag{5.133}$$

Proof Fix any $\lambda_0 \in (0, \lambda_1(B_1))$ and consider $\lambda \in [0, \lambda_0]$ only. For any $\lambda, \mu \in [0, \lambda_0]$, $\mu < \lambda$, we see from (5.132) that

$$\tau_{\mu,u_\lambda}^{2^*-2} = \frac{\int_{B_1}(|\nabla U_\lambda|^2 - \mu|U_\lambda|^2)\,dx}{\int_{B_1}|U_\lambda|^{2^*}\,dx} = 1 + (\lambda - \mu)\beta(\lambda),$$

where

$$\beta(\lambda) = \frac{\int_{B_1}|U_\lambda|^2\,dx}{\int_{B_1}(|\nabla U_\lambda|^2 - \lambda|U_\lambda|^2)\,dx} \le \frac{1}{\lambda_1(B_1) - \lambda}.$$

Thus, we derive from (5.130) and the mean value theorem that

$$0 < e(\mu) - e(\lambda) \le \max_{t>0} I_\mu(tu_\lambda) - e(\lambda) = \tau_{\mu,u_\lambda}^{2^*}e(\lambda) - e(\lambda)$$

$$= \left[1 + (\lambda - \mu)\beta(\lambda)\right]^{N/2}e(\lambda) - e(\lambda)$$

$$\le \frac{N}{2}\left[1 + \beta(\lambda)(\lambda - \mu)\right]^{\frac{N-2}{2}}\beta(\lambda)e(\lambda)(\lambda - \mu)$$

$$\le \frac{N}{2}\left(1 + \frac{\lambda - \mu}{\lambda_1(B_1) - \lambda}\right)^{\frac{N-2}{2}}\frac{1}{\lambda_1(B_1) - \lambda}e(\lambda)(\lambda - \mu)$$

$$\le \frac{N}{2}\left(\frac{\lambda_1(B_1)}{\lambda_1(B_1) - \lambda_0}\right)^{\frac{N-2}{2}}\frac{1}{\lambda_1(B_1) - \lambda_0}e(0)(\lambda - \mu) = C(\lambda_0)(\lambda - \mu).$$

Similarly, for any $\lambda, \mu \in [0, \lambda_0]$ with $\mu > \lambda$, we also have $0 < e(\lambda) - e(\mu) \le C(\lambda_0)(\mu - \lambda)$. Hence, for any $\lambda, \mu \in [0, \lambda_0]$, we have $|e(\mu) - e(\lambda)| \le C(\lambda_0)|\mu - \lambda|$. That is, $\lambda \mapsto e(\lambda)$ is Lipschitz continuous with respect to $\lambda \in [0, \lambda_0]$. In particular, it is absolutely continuous and so it follows from Lemma 5.17 that (5.133) holds for all $\lambda \in [0, \lambda_0]$. $\qquad\square$

Now, we fix any $\lambda > 0$ and consider Eq. (5.124). Clearly, there exists $R_0 > 0$ such that $\lambda = \lambda_1(B_{R_0})$. Then by Brezis and Nirenberg [19], we know that (5.124) has a least energy solution u_R with least energy $c(R) := I_{B_R}(u_R)$ for any $R \in (0, R_0)$, where $I_{B_R} := I_{\lambda, B_R}$ is the corresponding functional of (5.11). By Gidas et al. [53], we know that u_R must be radial, and we write $u_R(r) = u_R(x)$ for convenience, where $r = |x|$. Recalling $u_R \in C^2([0, R])$, we denote $u_R'(R) := \frac{d}{dr}u_R(r)|_{r=R}$. Then we have the following lemma, which plays a crucial role in the proof of Theorem 5.9.

Lemma 5.19 *$c(R)$ is strictly decreasing with respect to R, and for any $R \in (0, R_0)$, there holds*

$$c(R) = I_{B_R}(u_R) = \frac{1}{N}S^{N/2} - \frac{|S^{N-1}|}{2}\int_0^R |u_r'(r)|^2 r^{N-1}\,dr.$$

Here, S^{N-1} is the unit sphere of $\mathbb{R}^N$ and $|S^{N-1}|$ is the Lebesgue measure of S^{N-1}.

Proof By the Pohozaev identity, we have

$$\lambda \int_{B_R} |u_R|^2 \, dx = \frac{1}{2} \int_{\partial B_R} (x \cdot \nu) \left(\frac{\partial u_R}{\partial \nu} \right)^2 \, d\sigma = \frac{|S^{N-1}|}{2} |u_R'(R)|^2 R^N. \qquad (5.134)$$

Let $v_R(x) := R^{\frac{N-2}{2}} u_R(Rx)$, then v_R satisfies $-\Delta v = \lambda R^2 v + v^{2^*-1}$ in B_1. That is, v_R is a least energy solution of (5.129) with λR^2. Recalling that $U_{\lambda R^2}$ is also a least energy solution of (5.129) with λR^2, we derive from Lemma 5.17 that

$$\int_{B_1} |v_R|^2 \, dx = \int_{B_1} |U_{\lambda R^2}|^2 \, dx = -2e'(\lambda R^2) \qquad (5.135)$$

holds for almost every $R \in (0, R_0)$. It is easily seen that $c(R) = I_{B_R}(u_R) = I_{\lambda R^2}(v_R) = e(\lambda R^2)$, which implies that $c(R)$ is strictly decreasing with respect to R. From (5.134), we have

$$\int_{B_1} |v_R|^2 \, dx = R^{-2} \int_{B_R} |u_R|^2 \, dx = \frac{|S^{N-1}|}{2\lambda} |u_R'(R)|^2 R^{N-2}.$$

Combining this with Lemma 5.18 and (5.135), we have

$$c(R) = e(\lambda R^2) = e(0) - \frac{1}{2} \int_0^{\lambda R^2} \left(\int_{B_1} |U_\mu|^2 \, dx \right) d\mu$$

$$= e(0) - \frac{1}{2} \int_0^R \left(\int_{B_1} |v_r|^2 \, dx \right) 2\lambda r \, dr \quad (\text{let } \mu = \lambda r^2)$$

$$= \frac{1}{N} S^{N/2} - \frac{|S^{N-1}|}{2} \int_0^R |u_r'(r)|^2 r^{N-1} \, dr.$$

This completes the proof. $\qquad \qquad \square$

Recall the fact that

$$\lim_{R \to 0} u_R(0) = +\infty. \qquad (5.136)$$

In fact, if there exists $R_n \to 0$ such that $u_{R_n}(0) \leq M$ for some constant $M > 0$. Then by the dominated convergence theorem, we have

$$0 < Ne(0) = \lim_{R_n \to 0} Nc(R_n) = \lim_{R_n \to 0} \int_{B_{R_n}} |u_{R_n}|^{2^*} \, dx = 0,$$

which yields a contradiction. Note that $u_R(r)$ is the unique solution of

$$\begin{cases} -u'' - \frac{N-1}{r}u' = \lambda u + u^{2^*-1}, \\ u'(0) = 0, \quad u(0) = u_R(0). \end{cases} \tag{5.137}$$

As in [11], we define

$$y(t) := \lambda^{-\frac{N-2}{4}} u\left((N-2)\lambda^{-\frac{1}{2}}t^{-\frac{1}{N-2}}\right), \tag{5.138}$$

then equation $-u'' - \frac{N-1}{r}u' = \lambda u + u^{2^*-1}$ becomes

$$y''(t) + t^{-\frac{2(N-1)}{N-2}}\left(y(t) + y^{2^*-1}(t)\right) = 0.$$

Denoting $k = \frac{2(N-1)}{N-2} > 2$, we have $2^* - 1 = 2k - 3$. Consider

$$\begin{cases} y''(t) + t^{-k}\left(y(t) + y^{2k-3}(t)\right) = 0, \\ \lim_{t \to +\infty} y(t) = \gamma > 0, \end{cases} \tag{5.139}$$

Since $k > 2$, the existence of a unique positive solution $y_\gamma(t)$ for (5.139) is ensured for t large (cf. [11]). Define

$$T(\gamma) := \inf\{t > 0 \mid y_\gamma(\tau) > 0 \; \forall \tau > t\}, \tag{5.140}$$

and

$$z_\gamma(t) := \gamma t\left(t^{k-2} + \frac{1 + \gamma^{2k-4}}{k-1}\right)^{-\frac{1}{k-2}}. \tag{5.141}$$

Then, (see Lemma 1 and Remark 1 in [11]) $y_\gamma(t) \le z_\gamma(t)$ for any $t \ge T(\gamma)$ and $z_\gamma(t)$ satisfies

$$z''(t) + t^{-k}\frac{1 + \gamma^{2k-4}}{\gamma^{2k-4}}z^{2k-3}(t) = 0. \tag{5.142}$$

We recall some results from [11].

Lemma 5.20 ([11, Theorem 3]) *Suppose $k > 2$. Then there exists positive constants $C_1, C_2,$ and γ_0, which depend on k, such that the following inequalities hold for all $\gamma \ge \gamma_0$.*

(1) *If $k = 3$, then $C_1 \log \gamma < T(\gamma) < C_2 \log \gamma$.*
(2) *If $k < 3$, then $C_1\gamma^{6-2k} < T(\gamma) < C_2\gamma^{6-2k}$.*

Using Lemma 5.20, we can prove the following lemma.

Lemma 5.21 *Suppose* $2 < k \leq 3$. *Then there exists positive constants* C_3 *and* $\gamma_1 \geq \gamma_0$, *which depend on* k, *such that*

$$y'_\gamma(T(\gamma)) := \frac{d}{dt} y_\gamma(t)|_{t=T(\gamma)} \leq \frac{C_3}{\gamma}$$

holds for all $\gamma \geq \gamma_1$.

Proof From (5.141), we have $z_\gamma(t) \leq (k-1)^{\frac{1}{k-2}} \frac{t}{\gamma}$ and

$$z'_\gamma(t) = \gamma \frac{1 + \gamma^{2k-4}}{k-1} \left(t^{k-2} + \frac{1+\gamma^{2k-4}}{k-1} \right)^{-\frac{1}{k-2}-1} < (k-1)^{\frac{1}{k-2}} \frac{1}{\gamma}$$

for all $t \geq T(\gamma)$. Combining these with (5.139), (5.142), we have

$$y'_\gamma(T(\gamma)) := \frac{d}{dt} y_\gamma(t)|_{t=T(\gamma)} = \int_{T(\gamma)}^{+\infty} \frac{y_\gamma(t) + y_\gamma^{2k-3}(t)}{t^k} \, dt$$

$$\leq \int_{T(\gamma)}^{+\infty} \frac{z_\gamma(t)}{t^k} \, dt + \int_{T(\gamma)}^{+\infty} \frac{z_\gamma^{2k-3}(t)}{t^k} \, dt$$

$$\leq (k-1)^{\frac{1}{k-2}} \frac{1}{\gamma} \int_{T(\gamma)}^{+\infty} t^{1-k} \, dt - \frac{\gamma^{2k-4}}{1+\gamma^{2k-4}} \int_{T(\gamma)}^{+\infty} z''_\gamma(t) \, dt$$

$$= (k-1)^{\frac{1}{k-2}} \frac{1}{k-2} T(\gamma)^{2-k} \frac{1}{\gamma} + \frac{\gamma^{2k-4}}{1+\gamma^{2k-4}} z'_\gamma(T(\gamma))$$

$$\leq \left(\frac{(k-1)^{\frac{1}{k-2}}}{k-2} T(\gamma)^{2-k} + (k-1)^{\frac{1}{k-2}} \right) \frac{1}{\gamma}.$$

Noting from Lemma 5.20 that $T(\gamma) \to +\infty$ as $\gamma \to +\infty$, we conclude the proof. $\qquad\square$

Lemma 5.22 *There exists* $R_1 \in (0, R_0)$ *and positive constants* C_4, C_5, *which depend only on* N *and* λ, *such that for any* $R \leq R_1$ *there holds*

$$\frac{C_4}{u_R(0)R^{N-1}} \leq |u'_R(R)| \leq \frac{C_5}{u_R(0)R^{N-1}}. \tag{5.143}$$

Proof Denote $y_R(t) := \lambda^{-\frac{N-2}{4}} u_R \left((N-2)\lambda^{-\frac{1}{2}} t^{-\frac{1}{N-2}} \right)$ in (5.138). Then y_R satisfies (5.139) with $k = \frac{2(N-1)}{N-2}$ and

$$\gamma = \lambda^{-\frac{N-2}{4}} u_R(0), \quad T(\gamma) = (N-2)^{N-2}\lambda^{-\frac{N-2}{2}} R^{2-N}, \tag{5.144}$$

and so

$$y'_R(T(\gamma)) = -\lambda^{-\frac{N}{4}} u'_R(R) T(\gamma)^{-\frac{N-1}{N-2}} = (N-2)^{1-N}\lambda^{\frac{N-2}{4}} |u'_R(R)| R^{N-1}.$$

Then by (5.136) and Lemma 5.21, there exists $R_1 \in (0, R_0)$ and constants C_5, which depend only on N and λ, such that for any $R \le R_1$ the right-hand side of (5.143) holds. On the other hand, from (5.137) and Lemma 5.19 we have

$$R^{N-1}|u'_R(R)| = \int_0^R r^{N-1}\left(\lambda u_R(r) + u_R^{2^*-1}(r)\right) dr$$

$$> \frac{1}{u_R(0)} \int_0^R r^{N-1} u_R^{2^*}(r)\, dr = \frac{1}{u_R(0)} \frac{1}{|S^{N-1}|} \int_{B_R} u_R^{2^*}(x)\, dx$$

$$= \frac{Nc(R)}{|S^{N-1}|} \frac{1}{u_R(0)} \ge \frac{Nc(R_1)}{|S^{N-1}|} \frac{1}{u_R(0)} = \frac{C_4}{u_R(0)},$$

for $R \le R_1$, so the left-hand side of (5.143) holds. $\qquad\square$

Lemma 5.23 *There exists $R_2 \in (0, R_1]$ and positive constants C_6, C_7, C_8, C_9, which depend only on N and λ, such that the following inequalities hold for any $R \le R_2$.*

(1) *If $N = 4$, then*

$$\lambda^{1/2} e^{C_6 R^{-2}} < u_R(0) < \lambda^{1/2} e^{C_7 R^{-2}}.$$

(2) *If $N \ge 5$, then*

$$C_8 R^{-\frac{(N-2)^2}{2N-8}} < u_R(0) < C_9 R^{-\frac{(N-2)^2}{2N-8}}.$$

Proof Noting that $k = \frac{2(N-1)}{N-2} = 3$ when $N = 4$, $k < 3$ when $N \ge 5$, this lemma follows directly from Lemma 5.20 and (5.144). $\qquad\square$

Now we can finish the proof of Theorem 5.9.

Proof (Proof of Theorem 5.9) Fix any $R \in (0, R_2]$. First, we assume $N \ge 5$. Then we see from Lemmas 5.22 and 5.23 that

$$\int_0^R |u_r'(r)|^2 r^{N-1}\, dr \geq C_4^2 \int_0^R \frac{1}{|u_r(0)|^2 r^{N-1}}\, dr$$

$$\geq C(N, \lambda) \int_0^R r^{\frac{(N-2)^2}{N-4} - N + 1}\, dr$$

$$\geq C_1(N, \lambda) R^{\frac{2N-4}{N-4}}.$$

Similarly,

$$\int_0^R |u_r'(r)|^2 r^{N-1}\, dr \leq C_2(N, \lambda) R^{\frac{2N-4}{N-4}}.$$

Thus by Lemma 5.19, we see that (5.126) in Theorem 5.9 holds. Now, we assume $N = 4$, then

$$\int_0^R |u_r'(r)|^2 r^{N-1}\, dr \geq C_4^2 \int_0^R \frac{1}{|u_r(0)|^2 r^3}\, dr \geq \frac{C_4^2}{\lambda} \int_0^R e^{-2C_7 r^{-2}} r^{-3}\, dr$$

$$= \frac{C_4^2}{2\lambda} \int_{R^{-2}}^{+\infty} e^{-2C_7 t}\, dt = \frac{C_4^2}{4\lambda C_7} e^{-2C_7 R^{-2}}.$$

Similarly,

$$\int_0^R |u_r'(r)|^2 r^{N-1}\, dr \leq \frac{C_5^2}{4\lambda C_6} e^{-2C_6 R^{-2}}.$$

namely (5.127) holds. This completes the proof. $\square$

Chapter 6
A Linearly Coupled Schrödinger System with Critical Exponent

Abstract As introduced in Chap. 1, we consider the linearly coupled system (1.9) with $1 < p < q = 2^* - 1$. We make a systematical study of the ground state solutions, including existence, nonexistence, and asymptotic behaviors. Moreover, our result on the parameter range for the existence and nonexistence is almost *optimal*. Our proof is purely variational, and the key step is to establish accurate upper bounds of the least energy.

6.1 Main Results

Consider the following linearly coupled Schrödinger equations:

$$\begin{cases} -\Delta u + \mu u = |u|^{p-1}u + \lambda v, & x \in \mathbb{R}^N, \\ -\Delta v + \nu v = |v|^{q-1}v + \lambda u, & x \in \mathbb{R}^N, \\ u(x), v(x) \to 0 & \text{as } |x| \to +\infty, \end{cases} \qquad (6.1)$$

where μ, ν, and λ are all positive parameters and $p, q > 1$, and $p, q \leq 2^* - 1$ if $N \geq 3$. These kind of systems arise as mathematical models from nonlinear optics [4]. There is an obvious different property for system (6.1) compared to BEC type systems that are studied in the previous four chapters: system (6.1) has *no* semi-trivial solutions provided $\lambda \neq 0$ because of the linear coupling, namely (6.1) is a fully coupled system.

Let us recall some previous results first. In the case where $N \leq 3$, $\mu = \nu = 1$ and $p = q = 3$, Ambrosetti et al. [8] proved that (6.1) has multi-bump solitons for $\lambda > 0$ small enough. When the nonlinearities $|u|^{p-1}u$ and $|v|^{q-1}v$ are replaced by more general ones $f(x, u) = (1 + c(x))|u|^{p-1}u$ and $g(x, v) = (1 + d(x))|v|^{p-1}v$, respectively, system (6.1) has been studied by Ambrosetti [5] for dimension $N = 1$ and by Ambrosetti et al. [9] for dimensions $N \geq 2$. In particular, when $\mu = \nu = 1$, $1 < p = q < 2^* - 1$ and $0 < \lambda < 1$, Ambrosetti et al. proved that system (6.1) has a positive ground state solution (see [9, Sect. 3]). We note that this result can be extended to the more general case where $0 < \lambda < \sqrt{\mu\nu}$ and $1 < p, q < 2^* - 1$ via

© Springer-Verlag Berlin Heidelberg 2015
Z. Chen, *Solutions of Nonlinear Schrödinger Systems*, Springer Theses,
DOI 10.1007/978-3-662-45478-7_6

the classical result of Brezis and Lieb [18]. Indeed, Brezis and Lieb [18] consider some systems of equations

$$- \Delta u_i(x) = g^i(u(x)), \ i = 1, \ldots, n, \tag{6.2}$$

where the n functions $g^i : \mathbb{R}^n \to \mathbb{R}$ are the gradients of some function $G \in C^1(\mathbb{R}^n)$, namely $g^i(u) = \partial G(u)/\partial u_i$. Under some conditions on g^i (see (2.2), (2.3), (2.4), and (2.8) in [18]), they proved that (6.2) has a ground state solution which belongs to $H^1(\mathbb{R}^N) \cap W_{loc}^{2,q}(\mathbb{R}^N)$ for any $q < +\infty$ (see [18, Theorems 2.2 and 2.3]). Recently, Byeon et al. [21] proved that ground state solutions of (6.2) obtained in [18] must be radially symmetric up to a translation. Remark that system (6.1) is a special case of (6.2), and assumptions $0 < \lambda < \sqrt{\mu\nu}$ with $1 < p, q < 2^* - 1$ are consistent with those conditions on g^i in [18]; it follows that (6.1) possesses a ground state solution in this case. Recently, a more general version of (6.1), namely

$$\begin{cases} -\Delta u + u = f(u) + \lambda v, & x \in \mathbb{R}^N, \\ -\Delta v + v = g(v) + \lambda u, & x \in \mathbb{R}^N, \\ u(x), v(x) \to 0, & \text{as } |x| \to +\infty, \end{cases} \tag{6.3}$$

has been studied in [31, 37]. Here f, g are general Caratheodory functions with subcritical growth. In particular, under almost *optimal* conditions on nonlinearities f and g, we proved in [37] that (6.3) has at least *two positive* solutions (one is ground state but the other one is not) for $\lambda > 0$ small. See also [40] for the corresponding semiclassical states of (6.3).

Remark that all the papers mentioned above only deal with the subcritical case $1 < p, q < 2^* - 1$. To the best of our knowledge, the critical case for such a linearly coupled system has *not* ever been studied in the literature. In this chapter, we study the critical case, namely $1 < p \leq q = 2^* - 1$. Since we are concerned with the ground state solutions, we assume

$$0 < \lambda < \sqrt{\mu\nu}$$

in the sequel. This assumption is needed to guarantee that the Nehari manifold is bounded away from $(0, 0)$. Remark that, if $p = q = 2^* - 1$, then system (6.1) has no nontrivial solutions by the Pohozaev identity. Therefore, here we only consider the case $1 < p < q = 2^* - 1$, and then system (6.1) turns to be

$$\begin{cases} -\Delta u + \mu u = |u|^{p-1}u + \lambda v, & x \in \mathbb{R}^N, \\ -\Delta v + \nu v = |v|^{2^*-2}v + \lambda u, & x \in \mathbb{R}^N, \\ u(x), v(x) \to 0 & \text{as } |x| \to +\infty. \end{cases} \tag{6.4}$$

Remark that, although system (6.4) here and system (5.1) studied in Chaps. 4 and 5 are both two-coupled critical systems, as we will see in the following, the study of system (6.4) is actually quite *different* from system (5.1). One reason is that system (6.4) is *linearly* coupled while system (5.1) is *nonlinearly* coupled.

Before giving our results, we give some notations first. As in Chap. 2, we denote the norm of $L^p(\mathbb{R}^N)$ by $|u|_p = (\int_{\mathbb{R}^N} |u|^p \, dx)^{\frac{1}{p}}$. Denote the standard scalar product and norm of $H^1(\mathbb{R}^N)$ by

$$\langle u, v \rangle := \int_{\mathbb{R}^N} (\nabla u \nabla v + uv) \, dx, \quad \|u\|^2 := \langle u, u \rangle.$$

Define $H := H^1(\mathbb{R}^N) \times H^1(\mathbb{R}^N)$ with norm $\|(u, v)\|^2 := \|u\|^2 + \|v\|^2$. For convenience, we denote

$$\|u\|_\mu^2 := \int_{\mathbb{R}^N} \left(|\nabla u|^2 + \mu |u|^2 \right) dx.$$

Clearly, we can look for solutions of system (6.4) by seeking critical points of a C^2 functional $I_{\mu,\nu,\lambda} : H \to \mathbb{R}$, where

$$I(u, v) := I_\lambda(u, v) := I_{\mu,\nu,\lambda}(u, v) := \frac{1}{2} \|u\|_\mu^2 + \frac{1}{2} \|v\|_\nu^2$$
$$- \frac{1}{p+1} |u|_{p+1}^{p+1} - \frac{1}{2^*} |v|_{2^*}^{2^*} - \lambda \int_{\mathbb{R}^N} uv \, dx.$$

In the following, we will omit the subscripts μ, ν, λ when there is no confusion arising. To study the ground state solutions, we define the Nehari manifold as

$$M := M_\lambda := M_{\mu,\nu,\lambda} := \left\{ (u, v) \in H \backslash \{(0, 0)\} \ : \ I'(u, v)(u, v) = 0 \right\}.$$

Since $\mu, \nu > 0$ and $0 < \lambda < \sqrt{\mu\nu}$, it is easy to prove the existence of constant $\rho_{\mu,\nu,\lambda} > 0$ such that

$$\|u\|^2 + \|v\|^2 \geq \rho_{\mu,\nu,\lambda}, \quad \forall (u, v) \in M_{\mu,\nu,\lambda}.$$

By this we easily conclude that M is a complete smooth manifold, and critical points of I constrained on M are also critical points of I on H and so solutions of (6.4). Define the least energy

$$m_{\mu,\nu,\lambda} := \inf_{(u,v) \in M_{\mu,\nu,\lambda}} I(u, v). \tag{6.5}$$

We call that a solution $(u, v) \in M_{\mu,\nu,\lambda}$ is a *ground state* solution of system (6.4), if $I(u, v) = m_{\mu,\nu,\lambda}$.

Assume that C_{p+1} is the sharp constant of Sobolev embedding $H^1(\mathbb{R}^N) \hookrightarrow L^{p+1}(\mathbb{R}^N)$

$$\int_{\mathbb{R}^N} |\nabla u|^2 + |u|^2 \, dx \geq C_{p+1} \left(\int_{\mathbb{R}^N} |u|^{p+1} \, dx \right)^{\frac{2}{p+1}} , \tag{6.6}$$

and S is the sharp constant of $D^{1,2}(\mathbb{R}^N) \hookrightarrow L^{2^*}(\mathbb{R}^N)$

$$\int_{\mathbb{R}^N} |\nabla u|^2 \, dx \geq S \left(\int_{\mathbb{R}^N} |u|^{2^*} \, dx \right)^{\frac{2}{2^*}} . \tag{6.7}$$

As before, the norm of $D^{1,2}(\mathbb{R}^N)$ is

$$\|u\|_{D^{1,2}} := \left(\int_{\mathbb{R}^N} |\nabla u|^2 \, dx \right)^{1/2} .$$

Define a constant

$$\mu_0 := \left[\frac{2(p+1)}{N(p-1)} S^{\frac{N}{2}} C_{p+1}^{-\frac{p+1}{p-1}} \right]^{(\frac{p+1}{p-1} - \frac{N}{2})-1} . \tag{6.8}$$

Then our main result in this chapter is following.

Theorem 6.1 *Assume $N \geq 3$, $1 < p < 2^* - 1$ and $\mu, \nu > 0, 0 < \lambda < \sqrt{\mu\nu}$. Let μ_0 be in (6.8).*

(1) *If $0 < \mu \leq \mu_0$, then system (6.4) has a positive ground state solution (u, v), such that $u, v \in C^2(\mathbb{R}^N, \mathbb{R})$ are both radial symmetric decreasing.*
(2) *If $\mu > \mu_0$, then there exists $\lambda_{\mu,\nu} \in [\sqrt{(\mu - \mu_0)\nu}, \sqrt{\mu\nu})$ such that*

   (i) *if $\lambda < \lambda_{\mu,\nu}$, then system (6.4) has no ground state solutions.*
   (ii) *if $\lambda > \lambda_{\mu,\nu}$, then system (6.4) has a positive ground state solution (u, v), such that $u, v \in C^2(\mathbb{R}^N, \mathbb{R})$ are both radial symmetric decreasing.*

Remark 6.1 It is interesting that, whether the ground state solution of (6.4) exists or not depends heavily on the relation of μ, ν, λ and μ_0, and μ_0 can be seen as a critical value for the existence of ground state solutions. In particular, the case $\mu > \mu_0$ is more delicate, and $\lambda_{\mu,\nu}$ can be seen as a critical value in this case. The existence of the ground state solutions for $\lambda = \lambda_{\mu,\nu}$ seems much more difficult and remains an open question (see Remark 6.4). Therefore, the ranges of parameter λ for the existence of ground state solutions in Theorem 6.1 is almost optimal.

Remark 6.2 Though the exact value of C_{p+1} seems unknown, we have the following lower bound estimates for C_{p+1}:

$$C_{p+1} > \widetilde{C} := \alpha^{-\alpha}(1-\alpha)^{-(1-\alpha)} S^{1-\alpha}, \qquad (6.9)$$

where

$$\alpha := N\left(\frac{1}{p+1} - \frac{1}{2^*}\right) \in (0, 1).$$

Using (6.8) and (6.9) we can prove that

$$\mu_0 < \overline{\mu}_0 := \alpha(1-\alpha)^{\frac{N-2}{2}(\frac{p+1}{p-1} - \frac{N}{2})^{-1}} < 1, \qquad (6.10)$$

see Lemma 6.3.

Remark 6.3 We can also give a estimate from above for $\lambda_{\mu,\nu}$, that is, some number $\tilde{\lambda}_{\mu,\nu} < \sqrt{\mu\nu}$ such that $\lambda_{\mu,\nu} < \tilde{\lambda}_{\mu,\nu}$, see Lemma 6.4.

Now, we turn to study some further properties of the ground state solutions obtained in Theorem 6.1. Precisely, we have the following result.

Theorem 6.2 *Assume that μ, ν, λ satisfy the hypotheses in (1) or (2)-(ii) of Theorem 6.1. Let (u, v) be any a ground state solution of (6.4) which exists by Theorem 6.1, then up to a translation, $u, v \in C^2(\mathbb{R}^N, \mathbb{R})$ are positive radial symmetric decreasing. Moreover, there exists a positive constant $C = C(\mu, \nu, \lambda)$ independent of (u, v) such that*

$$\|u\|_{L^\infty(\mathbb{R}^N)} + \|v\|_{L^\infty(\mathbb{R}^N)} \le C.$$

Finally, fix any $\mu \in (0, \mu_0)$, $\nu > 0$ and let $\lambda_n \in (0, \sqrt{\mu\nu})$ such that $\lambda_n \to 0$ as $n \to +\infty$. Let $(u_{\lambda_n}, v_{\lambda_n})$ be any positive radial ground state solution of (6.4) with $\lambda = \lambda_n$. Then, passing to a subsequence, $(u_{\lambda_n}, v_{\lambda_n}) \to (u_0, 0)$ strongly in H, where u_0 is a positive radial ground state solution of

$$-\Delta u + \mu u = |u|^{p-1}u, \quad u \in H^1(\mathbb{R}^N).$$

Results in this chapter were published in a joint work with Zou [34]. Later, in a subsequent work [38], we applied Theorem 6.1 to the corresponding singularly perturbed problem:

$$\begin{cases} -\varepsilon^2 \Delta u + a(x)u = u^p + \lambda v, & x \in \mathbb{R}^N, \\ -\varepsilon^2 \Delta v + b(x)v = v^{2^*-1} + \lambda u, & x \in \mathbb{R}^N, \\ u, v > 0 \text{ in } \mathbb{R}^N, \quad u(x), v(x) \to 0 \text{ as } |x| \to \infty, \end{cases} \qquad (6.11)$$

where $\varepsilon > 0$ is a small parameter and a, b are both continuous potentials with positive lower bounds. Remark that, as far as the *semiclassical states* related to system (6.4) are concerned, we are naturally led to study system (6.11). Under some additional assumptions on $a(x)$ and λ, for ε sufficiently small, we proved that (6.11) has a positive solution, which concentrates to the ground state solutions of system (6.4) as $\varepsilon \to 0$. The interesting thing is that we do *not* need any further assumptions on the potential $b(x)$. See [38] for details. The so-called critical frequency case $\inf_{x \in \mathbb{R}^N} \min\{a(x), b(x)\} = 0$ seems more tough and remains open.

We will prove Theorem 6.1 in Sect. 6.2, and the proof turns out to be very technical. Theorem 6.2 will be proved in Sect. 6.3.

6.2 Sharp Parameter Ranges

In the sequel we always assume that $0 < \lambda < \sqrt{\mu\nu}$. In this section, we give the proof of Theorem 6.1. For any $(u, v) \in H \backslash \{(0, 0)\}$, we have

$$
\max_{t>0} I_\lambda(tu, tv) = I_\lambda(t_{\lambda,u,v}u, t_{\lambda,u,v}v)
$$
$$
= \left(\frac{1}{2} - \frac{1}{p+1} \right) t_{\lambda,u,v}^{p+1} |u|_{p+1}^{p+1} + \left(\frac{1}{2} - \frac{1}{2^*} \right) t_{\lambda,u,v}^{2^*} |v|_{2^*}^{2^*}, \quad (6.12)
$$

where $t_{\lambda,u,v} > 0$ satisfies $\varphi(\lambda, u, v, t_{\lambda,u,v}) = 0$, and φ is defined as

$$
\varphi(\lambda, u, v, t) := \|u\|_\mu^2 + \|v\|_\nu^2 - 2\lambda \int_{\mathbb{R}^N} uv\, dx - t^{p-1}|u|_{p+1}^{p+1} - t^{2^*-2}|v|_{2^*}^{2^*}. \quad (6.13)
$$

This implies $(t_{\lambda,u,v}u, t_{\lambda,u,v}v) \in M_\lambda$, so

$$
m_{\mu,\nu,\lambda} = \inf_{(u,v) \in H \backslash \{(0,0)\}} \max_{t>0} I(tu, tv). \quad (6.14)
$$

Since $\varphi(\lambda, u, v, t)$ is decreasing with respect to $t > 0$ and $\varphi(\lambda, u, v, 0) > 0$, so $t_{\lambda,u,v}$ is unique. Furthermore, $t_{\lambda,u,v} = 1$ for any $(u, v) \in M_\lambda$. Since

$$
\max_{t>0} I(tu, tv) \geq \max_{t>0} I(t|u|, t|v|), \quad (6.15)
$$

we also have

$$
m_{\mu,\nu,\lambda} = \inf_{(u,v) \in H \backslash \{(0,0)\}} \max_{t>0} I(t|u|, t|v|). \quad (6.16)
$$

Lemma 6.1 *For fixed $\mu, \nu > 0$, $m_{\mu,\nu,\lambda}$ is nonincreasing with respect to $\lambda > 0$.*

Proof Let $\lambda_1 < \lambda_2$. For any $(u, v) \in H \backslash \{(0, 0)\}$ and $t > 0$, we have $I_{\lambda_1}(t|u|, t|v|) \geq I_{\lambda_2}(t|u|, t|v|)$. Using (6.16) we get that $m_{\mu,\nu,\lambda_1} \geq m_{\mu,\nu,\lambda_2}$. $\qquad\square$

Now we define

$$f_{\beta,\gamma}(u) := \frac{1}{2}\|u\|_\beta^2 - \frac{1}{p+1}\gamma|u|_{p+1}^{p+1}, \quad g(v) := \frac{1}{2}\int_{\mathbb{R}^N} |\nabla v|^2\, dx - \frac{1}{2^*}|v|_{2^*}^{2^*}, \quad (6.17)$$

and denote $f_\beta := f_{\beta,1}$ for simplicity.

Lemma 6.2 *For any $(u, v) \in H$ with $u \neq 0$ and $v \neq 0$, there holds*

$$\max_{t>0} I(tu, tv) > \min\left\{\max_{t>0} f_{\mu-\lambda^2/\nu}(tu), \quad \max_{t>0} g(tv)\right\}.$$

Proof Fix any a pair $(u, v) \in H$ with $u \neq 0$ and $v \neq 0$. Since $2\lambda uv \leq \frac{\lambda^2}{\nu}u^2 + \nu v^2$, we have

$$I(tu, tv) \geq f_{\mu-\lambda^2/\nu}(tu) + g(tv).$$

Moreover, there exists $t_1, t_2 > 0$ such that

$$\max_{t>0} f_{\mu-\lambda^2/\nu}(tu) = f_{\mu-\lambda^2/\nu}(t_1 u), \quad \max_{t>0} g(tv) = g(t_2 v).$$

If $t_1 \geq t_2$, then $f_{\mu-\lambda^2/\nu}(t_2 u) > 0$ and so $I(t_2 u, t_2 v) > g(t_2 v) = \max_{t>0} g(tv)$. If $t_1 < t_2$, then $g(t_1 v) > 0$ and so $I(t_1 u, t_1 v) > f_{\mu-\lambda^2/\nu}(t_1 u) = \max_{t>0} f_{\mu-\lambda^2/\nu}(tu)$. □

Let w be the radially symmetric positive solution of $-\Delta u + u = u^p$, $u \in H^1(\mathbb{R}^N)$. By Kwong [60] it is known that w is unique up to a translation and attains the sharp constant C_{p+1} in (6.6), with energy

$$f_1(w) = \left(\frac{1}{2} - \frac{1}{p+1}\right) C_{p+1}^{\frac{p+1}{p-1}},$$

where f_1 is defined in (6.17). Therefore, $w_{\beta,\gamma}(x) := \beta^{\frac{1}{p-1}}\gamma^{-\frac{1}{p-1}}w(\sqrt{\beta}x)$ is the unique positive solution of

$$-\Delta u + \beta u = \gamma u^p, \quad u \in H^1(\mathbb{R}^N)$$

with energy

$$f_{\beta,\gamma}(w_{\beta,\gamma}) = \gamma^{-\frac{2}{p-1}}\beta^{\frac{p+1}{p-1} - \frac{N}{2}} f_1(w) = \left(\frac{1}{2} - \frac{1}{p+1}\right)\gamma^{-\frac{2}{p-1}}\beta^{\frac{p+1}{p-1} - \frac{N}{2}} C_{p+1}^{\frac{p+1}{p-1}}.$$

$$(6.18)$$

Here $\beta, \gamma > 0$. Denote $w_\beta := w_{\beta,1}$ for convenience. Define $\alpha := N(\frac{1}{p+1} - \frac{1}{2^*}) \in (0, 1)$, then

$$\frac{1}{p+1} = \frac{\alpha}{2} + \frac{1-\alpha}{2^*}.$$

Recalling $S, \mu_0, \overline{\mu}_0$ in (6.7), (6.8), and (6.10), we have the following lemma.

Lemma 6.3 *There holds* $0 < \mu_0 < \overline{\mu}_0 < 1$, *and*

$$f_\mu(w_\mu) \begin{cases} > \frac{1}{N}S^{N/2} & \text{if } \mu > \mu_0, \\ = \frac{1}{N}S^{N/2} & \text{if } \mu = \mu_0, \\ < \frac{1}{N}S^{N/2} & \text{if } \mu < \mu_0. \end{cases} \tag{6.19}$$

Proof By (6.8) and (6.18) we see that $f_{\mu_0}(w_{\mu_0}) = \frac{1}{N}S^{N/2}$. Recalling $p < 2^* - 1$, we have $\frac{p+1}{p-1} - \frac{N}{2} > 0$, thus (6.19) follows directly from (6.18). The fact $\overline{\mu}_0 < 1$ is guaranteed by $\alpha \in (0, 1)$ and the definition (6.10).

It suffices to prove $f_{\overline{\mu}_0}(w_{\overline{\mu}_0}) > \frac{1}{N}S^{N/2}$. For any $u \in H^1(\mathbb{R}^N)$, $u \neq 0$, we see from Hölder inequality and Young inequality that

$$\left(\int_{\mathbb{R}^N} |u|^{p+1}\, dx\right)^{\frac{2}{p+1}} \leq \left(\int_{\mathbb{R}^N} |u|^2\, dx\right)^\alpha \left(\int_{\mathbb{R}^N} |u|^{2^*}\, dx\right)^{\frac{2}{2^*}(1-\alpha)}$$

$$\leq \alpha\varepsilon^{1/\alpha}\int_{\mathbb{R}^N} |u|^2\, dx + (1-\alpha)\varepsilon^{-\frac{1}{1-\alpha}}\left(\int_{\mathbb{R}^N} |u|^{2^*}\, dx\right)^{\frac{2}{2^*}}.$$

Choose $C_0 > 0, \varepsilon_0 > 0$ such that

$$C_0\alpha\varepsilon_0^{1/\alpha} = 1, \quad C_0(1-\alpha)\varepsilon_0^{-\frac{1}{1-\alpha}} = S,$$

then we have

$$C_0 = S^{1-\alpha}(1-\alpha)^{-(1-\alpha)}\alpha^{-\alpha}$$

and

$$\|u\|_1^2 > \int_{\mathbb{R}^N} |u|^2\, dx + S\left(\int_{\mathbb{R}^N} |u|^{2^*}\, dx\right)^{\frac{2}{2^*}} \geq C_0\left(\int_{\mathbb{R}^N} |u|^{p+1}\, dx\right)^{\frac{2}{p+1}}.$$

This implies $C_{p+1} > C_0$ by letting $u = w$, namely (6.9) holds. Combining this with (6.18) we have

$$
f_\mu(w_\mu) > \left(\frac{1}{2} - \frac{1}{p+1}\right) \mu^{\frac{p+1}{p-1} - \frac{N}{2}} \left(S^{1-\alpha}(1-\alpha)^{-(1-\alpha)}\alpha^{-\alpha}\right)^{\frac{p+1}{p-1}}
$$
$$
= \frac{1}{N} S^{N/2}(1-\alpha)^{\frac{2-N}{2}} \alpha^{\frac{N}{2} - \frac{p+1}{p-1}} \mu^{\frac{p+1}{p-1} - \frac{N}{2}},
$$

which implies $f_{\overline{\mu}_0}(w_{\overline{\mu}_0}) > \frac{1}{N} S^{N/2}$. Therefore, $\mu_0 < \overline{\mu}_0$. $\qquad\square$

For any $\mu > \mu_0$, $v > 0$, we define a C^1 function $h_{\mu,v} : (0+\infty) \to \mathbb{R}$ by

$$
h_{\mu,v}(a) := \frac{\mu + va^2}{2a} - \frac{\mu_0}{2a}(1+a^2)^{-\frac{N}{2}(\frac{p+1}{p-1} - \frac{N}{2})^{-1}}. \tag{6.20}
$$

Then

$$
h_{\mu,v}(a) > \frac{\mu - \mu_0 + va^2}{2a} \geq \sqrt{(\mu - \mu_0)v},
$$

and so $h_{\mu,v}(a) \to +\infty$ as $a \to 0+$. On the other hand, we note that $h_{\mu,v}$ is increasing with respect to $a \in [\sqrt{\mu/v}, +\infty)$. Thus, there exists $a_{\mu,v} \in (0, \sqrt{\mu/v})$ such that

$$
\tilde{\lambda}_{\mu,v} := h_{\mu,v}(a_{\mu,v}) := \min_{a \in (0,+\infty)} h_{\mu,v}(a). \tag{6.21}
$$

Noting that $h_{\mu,v}(\sqrt{\mu/v}) < \sqrt{\mu v}$, we conclude

$$
\sqrt{(\mu - \mu_0)v} < \tilde{\lambda}_{\mu,v} < \sqrt{\mu v}. \tag{6.22}
$$

It is time for us to prove the following important result, where the least energy $m_{\mu,v,\lambda}$ is well investigated. It will play the *key* role in the proof of Theorem 6.1.

Lemma 6.4 (1) *If* $0 < \mu \leq \mu_0$, *then* $m_{\mu,v,\lambda} < \frac{1}{N} S^{N/2}$.

(2) *If* $\mu > \mu_0$, *then there exists some* $\lambda_{\mu,v} \in \left[\sqrt{(\mu - \mu_0)v}, \tilde{\lambda}_{\mu,v}\right)$, *here* $\tilde{\lambda}_{\mu,v}$ *is seen in* (6.21), *such that*

(i) *if* $0 < \lambda \leq \lambda_{\mu,v}$, *then* $m_{\mu,v,\lambda} = \frac{1}{N} S^{N/2}$;
(ii) *if* $\lambda_{\mu,v} < \lambda < \sqrt{\mu v}$, *then* $m_{\mu,v,\lambda} < \frac{1}{N} S^{N/2}$.

Proof (1) If $\mu \in (0, \mu_0)$, then it follows from Lemma 6.3 that

$$
\max_{t>0} I_{\mu,v,\lambda}(tw_\mu, 0) = \max_{t>0} f_\mu(tw_\mu) = f_\mu(w_\mu) < \frac{1}{N} S^{N/2},
$$

so $m_{\mu,v,\lambda} < \frac{1}{N} S^{N/2}$.

When $\mu = \mu_0$, we have $m_{\mu_0,\nu,\lambda} \leq f_{\mu_0}(w_{\mu_0}) = \frac{1}{N} S^{N/2}$. Assume by contradiction that $m_{\mu_0,\nu,\lambda} = \frac{1}{N} S^{N/2}$, then

$$I_{\mu_0,\nu,\lambda}(w_{\mu_0}, 0) = m_{\mu_0,\nu,\lambda}, \quad (w_{\mu_0}, 0) \in M_{\mu_0,\nu,\lambda},$$

which implies that $(w_{\mu_0}, 0)$ is a ground state solution of (6.4). Since $\lambda > 0$, we get that $w_{\mu_0} \equiv 0$, a contradiction. So $m_{\mu_0,\nu,\lambda} < \frac{1}{N} S^{N/2}$.

(2) We fix any $\mu > \mu_0$ and $\nu > 0$. First we claim that

$$m_{\mu,\nu,\lambda} = \frac{1}{N} S^{N/2} \quad \text{if } 0 < \lambda \leq \sqrt{(\mu - \mu_0)\nu}. \tag{6.23}$$

Assume $0 < \lambda \leq \sqrt{(\mu - \mu_0)\nu}$, then $\mu - \frac{\lambda^2}{\nu} \geq \mu_0$. Similarly as (6.14), we have

$$f_\mu(w_\mu) = \inf_{u \in H^1(\mathbb{R}^N) \backslash \{0\}} \max_{t > 0} f_\mu(tu).$$

By (6.7) we have

$$\inf_{v \in H^1(\mathbb{R}^N) \backslash \{0\}} \max_{t > 0} g(tv) = \frac{1}{N} S^{N/2}.$$

For any $(u, v) \in H \backslash \{(0, 0)\}$, if $v = 0$, then $\max\limits_{t>0} I(tu, 0) = \max\limits_{t>0} f_\mu(tu) \geq \frac{1}{N} S^{N/2}$. If $u = 0$, then $\max\limits_{t>0} I(0, tv) \geq \max\limits_{t>0} g(tv) \geq \frac{1}{N} S^{N/2}$. If $u \neq 0$ and $v \neq 0$, then by Lemmas 6.2 and 6.3 we have

$$\max_{t>0} I(tu, tv) > \min \left\{ \max_{t>0} f_{\mu-\lambda^2/\nu}(tu), \ \max_{t>0} g(tv) \right\} \geq \frac{1}{N} S^{N/2}. \tag{6.24}$$

Combining these with (6.14), we conclude $m_{\mu,\nu,\lambda} \geq \frac{1}{N} S^{N/2}$. On the other hand, since the equation $-\Delta v + \nu v = |v|^{2^*-2} v$, $v \in H^1(\mathbb{R}^N)$ has no nontrivial solutions by the Pohozaev identity, it is easily seen that S is also the sharp constant (although can not be attained) of

$$\int_{\mathbb{R}^N} |\nabla v|^2 + \nu |v|^2 \, dx \geq S \left(\int_{\mathbb{R}^N} |v|^{2^*} \, dx \right)^{\frac{2}{2^*}},$$

which implies that

$$m_{\mu,\nu,\lambda} \leq \inf_{v \in H^1(\mathbb{R}^N) \backslash \{0\}} \max_{t > 0} I(0, tv) = \frac{1}{N} S^{N/2}. \tag{6.25}$$

Consequently $m_{\mu,\nu,\lambda} = \frac{1}{N} S^{N/2}$, namely (6.23) holds.

To prove (i)–(ii), we let $0 < \lambda < \sqrt{\mu v}$. Recalling $a_{\mu,v}$ in (6.21), we define

$$\beta := \frac{\mu + v a_{\mu,v}^2 - 2\lambda a_{\mu,v}}{1 + a_{\mu,v}^2}, \quad \gamma := \frac{1}{1 + a_{\mu,v}^2}.$$

Clearly $\beta > 0, \gamma > 0$. It is easy to see from (6.18) that

$$
\begin{aligned}
m_{\mu,v,\lambda} &\leq \max_{t>0} I\left(t w_{\beta,\gamma}, t(a_{\mu,v} w_{\beta,\gamma})\right) \\
&< (1 + a_{\mu,v}^2) \max_{t>0} f_{\beta,\gamma}(t w_{\beta,\gamma}) = (1 + a_{\mu,v}^2) f_{\beta,\gamma}(w_{\beta,\gamma}) \\
&= (1 + a_{\mu,v}^2)^{\frac{N}{2}} (\mu + v a_{\mu,v}^2 - 2\lambda a_{\mu,v})^{\frac{p+1}{p-1} - \frac{N}{2}} \left(\frac{1}{2} - \frac{1}{p+1}\right) C_{p+1}^{\frac{p+1}{p-1}} =: A_0.
\end{aligned}
$$

By Lemma 6.3 we see that $A_0 \leq \frac{1}{N} S^{N/2}$ is equivalent to

$$(1 + a_{\mu,v}^2)^{\frac{N}{2}} (\mu + v a_{\mu,v}^2 - 2\lambda a_{\mu,v})^{\frac{p+1}{p-1} - \frac{N}{2}} \leq \mu_0^{\frac{p+1}{p-1} - \frac{N}{2}}.$$

By (6.20) and (6.21), we see that the above inequality is equivalent to $\lambda \geq \tilde{\lambda}_{\mu,v}$. Combining this with (6.22), we have $m_{\mu,v,\lambda} < \frac{1}{N} S^{N/2}$ for any $\lambda \in [\tilde{\lambda}_{\mu,v}, \sqrt{\mu v})$.
Define

$$\lambda_{\mu,v} := \inf\left\{\lambda < \sqrt{\mu v} \mid m_{\mu,v,\tau} < \frac{1}{N} S^{N/2}, \ \forall \tau \in [\lambda, \sqrt{\mu v})\right\}.$$

Then (6.23) implies $\lambda_{\mu,v} \in [\sqrt{(\mu - \mu_0)v}, \tilde{\lambda}_{\mu,v}]$, and for any $\lambda \in (\lambda_{\mu,v}, \sqrt{\mu v})$, there holds $m_{\mu,v,\lambda} < \frac{1}{N} S^{N/2}$, namely (ii) holds.

We claim that $m_{\mu,v,\lambda_{\mu,v}} = \frac{1}{N} S^{N/2}$, which implies $\lambda_{\mu,v} < \tilde{\lambda}_{\mu,v}$ immediately.

By (6.25) we have $m_{\mu,v,\lambda_{\mu,v}} \leq \frac{1}{N} S^{N/2}$. By the definition of $\lambda_{\mu,v}$, there exists $\lambda_n < \lambda_{\mu,v}$, $n \geq 1$ such that

$$\lim_{n \to +\infty} \lambda_n = \lambda_{\mu,v}, \quad m_n := m_{\mu,v,\lambda_n} \geq \frac{1}{N} S^{N/2}, \ \forall n \geq 1.$$

For any $(u, v) \in H \setminus \{(0,0)\}$, there exists $t_n > 0$ such that $\max_{t>0} I_{\mu,v,\lambda_n}(tu, tv) = I_{\mu,v,\lambda_n}(t_n u, t_n v)$. Since $\lambda_n \to \lambda_{\mu,v}$, we have $t_n \to t_0$ as $n \to +\infty$, where $t_0 > 0$ satisfies $\max_{t>0} I_{\mu,v,\lambda_{\mu,v}}(tu, tv) = I_{\mu,v,\lambda_{\mu,v}}(t_0 u, t_0 v)$. Then

$$\limsup_{n \to +\infty} m_n \leq \limsup_{n \to +\infty} I_{\mu,v,\lambda_n}(t_n u, t_n v) = I_{\mu,v,\lambda_{\mu,v}}(t_0 u, t_0 v).$$

This implies

$$\frac{1}{N}S^{N/2} \leq \limsup_{n\to+\infty} m_n \leq m_{\mu,\nu,\lambda_{\mu,\nu}},$$

and so $m_{\mu,\nu,\lambda_{\mu,\nu}} = \frac{1}{N}S^{N/2}$. By Lemma 6.1 and (6.23) we see that (i) holds. This completes the proof.   □

The following lemma is concerned with the nonexistence of ground state solutions.

Lemma 6.5 *If $\mu > \mu_0$ and $0 < \lambda < \lambda_{\mu,\nu}$, then problem (6.4) has no ground state solutions.*

Proof Fix any $\nu > 0$ and $\mu > \mu_0$. Assume by contradiction that there exists $\lambda \in (0, \lambda_{\mu,\nu})$ such that (6.4) has a ground state solution $(u_\lambda, v_\lambda) \neq (0, 0)$. Then $I_\lambda(u_\lambda, v_\lambda) = m_{\mu,\nu,\lambda} = \frac{1}{N}S^{N/2}$. By (6.4) we see that $u \neq 0$ and $v \neq 0$. By (6.15) and (6.16) we may assume that $u \geq 0$, $v \geq 0$ (or see the proof of Theorem 6.2). By standard elliptic regularity theory, we see that $u_\lambda, v_\lambda \in C^2(\mathbb{R}^N)$ and so $u_\lambda > 0$, $v_\lambda > 0$ via the strong maximum principle. Take $\lambda_1 \in (\lambda, \lambda_{\mu,\nu})$. Then we derive from Lemma 6.4, (6.12), and (6.14) that

$$\begin{aligned}
\frac{1}{N}S^{N/2} = m_{\mu,\nu,\lambda_1} &\leq \max_{t>0} I_{\lambda_1}(tu_\lambda, tv_\lambda) \\
&= I_{\lambda_1}(t_{\lambda_1,u_\lambda,v_\lambda}u_\lambda, t_{\lambda_1,u_\lambda,v_\lambda}v_\lambda) \\
&= I_\lambda(t_{\lambda_1,u_\lambda,v_\lambda}u_\lambda, t_{\lambda_1,u_\lambda,v_\lambda}v_\lambda) - (\lambda_1 - \lambda)t^2_{\lambda_1,u_\lambda,v_\lambda}\int_{\mathbb{R}^N} u_\lambda v_\lambda \, dx \\
&< I_\lambda(t_{\lambda_1,u_\lambda,v_\lambda}u_\lambda, t_{\lambda_1,u_\lambda,v_\lambda}v_\lambda) \leq I_\lambda(u_\lambda, v_\lambda) \\
&= \frac{1}{N}S^{N/2},
\end{aligned}$$

a contradiction. This completes the proof.   □

The following lemma is concerned with the existence of ground state solutions. The original idea is essentially due to Brezis and Nirenberg [19], and here we adapt it to study system (6.4).

Lemma 6.6 *Let $0 < \lambda < \sqrt{\mu\nu}$. If $m_{\mu,\nu,\lambda} < \frac{1}{N}S^{N/2}$, then system (6.4) admits a ground state solution $(u_0, v_0) \in C^2(\mathbb{R}^N, \mathbb{R})$ such that u_0, v_0 are both positive radial symmetric decreasing with respect to $r = |x| \in [0, +\infty)$.*

Proof Fix any $\mu, \nu, \lambda > 0$ with $0 < \lambda < \sqrt{\mu\nu}$, and denote $m := m_{\mu,\nu,\lambda} < \frac{1}{N}S^{N/2}$. Let $\varepsilon_n \in (0, 2^* - 1 - p)$ such that $\varepsilon_n \to 0$ as $n \to +\infty$. As pointed out in Sect. 6.1, by Brezis and Lieb [18] the following subcritical problem

$$\begin{cases}
-\Delta u + \mu u = |u|^{p-1}u + \lambda v, & x \in \mathbb{R}^N, \\
-\Delta v + \nu v = |v|^{2^*-2-\varepsilon_n}v + \lambda u, & x \in \mathbb{R}^N, \\
u(x), v(x) \to 0 & \text{as} |x| \to +\infty,
\end{cases} \tag{6.26}$$

has a ground state solution $(u_n, v_n) \in H$, with energy $c_n := J_n(u_n, v_n)$. Here

$$J_n(u, v) := \frac{1}{2}\|u\|_\mu^2 + \frac{1}{2}\|v\|_v^2 - \frac{1}{p+1}|u|_{p+1}^{p+1} - \frac{1}{2^* - \varepsilon_n}|v|_{2^*-\varepsilon_n}^{2^*-\varepsilon_n} - \lambda \int_{\mathbb{R}^N} uv\,dx.$$

By a similar proof of Theorem 1.3 in [31], we may assume that $u_n > 0, v_n > 0$, $u_n, v_n \in C^2(\mathbb{R}^N)$ and u_n, v_n are radial symmetric decreasing. Similarly as (6.14), we have

$$c_n = \inf_{(u,v)\in H\setminus\{(0,0)\}} \max_{t>0} J_n(tu, tv).$$

For any $(u, v) \in H\setminus\{(0,0)\}$, there exists $t_{u,v,n} > 0$ such that $\max_{t>0} J_n(tu, tv) = J_n(t_{u,v,n}u, t_{u,v,n}v)$. Recalling $t_{\lambda,u,v}$ in (6.12), it is easily seen that $t_{u,v,n} \to t_{\lambda,u,v}$ as $n \to +\infty$. Thus

$$\limsup_{n\to+\infty} c_n \leq \limsup_{n\to+\infty} J_n(t_{u,v,n}u, t_{u,v,n}v) = I(t_{\lambda,u,v}v, t_{\lambda,u,v}v) = \max_{t>0} I(tu, tv).$$

This implies $\limsup_{n\to+\infty} c_n \leq m$. So $\{c_n\}_{n\in\mathbb{N}}$ is bounded. Note that

$$c_n = J_n(u_n, v_n) - \frac{1}{p+1}J_n'(u_n, v_n)(u_n, v_n)$$

$$\geq \left(\frac{1}{2} - \frac{1}{p+1}\right)\left(\|u_n\|_\mu^2 + \|v_n\|_v^2 - 2\lambda \int_{\mathbb{R}^N} u_n v_n\,dx\right)$$

$$\geq C\left(\|u_n\|^2 + \|v_n\|^2\right), \tag{6.27}$$

we get that $\{(u_n, v_n)\}_{n\in\mathbb{N}}$ is bounded in H. Then passing to a subsequence, we may assume that $(u_n, v_n) \rightharpoonup (u_0, v_0)$ weakly in H, and so (u_0, v_0) satisfies (6.4). Since u_n, v_n are radial, we see that u_0, v_0 are radial and

$$\lim_{n\to+\infty} \int_{\mathbb{R}^N} |u_n|^{p+1}\,dx = \int_{\mathbb{R}^N} |u_0|^{p+1}\,dx. \tag{6.28}$$

Assume that $u_n(0) + v_n(0) = \max_{x\in\mathbb{R}^N} u_n(x) + \max_{x\in\mathbb{R}^N} v_n(x) \to +\infty$ as $n \to +\infty$. We will use a blowup analysis to get a contradiction. Define $K_n := \max\{u_n(0), v_n(0)\}$, then $K_n \to +\infty$. Define

$$U_n(x) = K_n^{-1}u_n(K_n^{-\alpha_n}x), \quad V_n(x) = K_n^{-1}v_n(K_n^{-\alpha_n}x), \quad \alpha_n = \frac{2^* - 2 - \varepsilon_n}{2}.$$

Then $1 = \max\{U_n(0), V_n(0)\} = \max\{\max\limits_{x \in \mathbb{R}^N} U_n(x), \max\limits_{x \in \mathbb{R}^N} V_n(x)\}$ and U_n, V_n satisfy

$$\begin{cases} -\Delta U_n + \mu K_n^{-2\alpha_n} U_n = K_n^{p-1-2\alpha_n} U_n^p + \lambda K_n^{-2\alpha_n} V_n, & x \in \mathbb{R}^N, \\ -\Delta V_n + \nu K_n^{-2\alpha_n} V_n = V_n^{2^*-1-\varepsilon_n} + \lambda K_n^{-2\alpha_n} U_n, & x \in \mathbb{R}^N. \end{cases}$$

Since

$$\int\limits_{\mathbb{R}^N} |\nabla U_n|^2 \, dx = K_n^{-\frac{(N-2)\varepsilon_n}{2}} \int\limits_{\mathbb{R}^N} |\nabla u_n|^2 \, dx,$$

we see that $\{(U_n, V_n)\}_{n \geq 1}$ is bounded in $D^{1,2}(\mathbb{R}^N) \times D^{1,2}(\mathbb{R}^N) =: D$. By elliptic estimates, for a subsequence we have $(U_n, V_n) \to (U, V) \in D$ uniformly in every compact subset of $\mathbb{R}^N$ as $n \to +\infty$, and U, V satisfy

$$-\Delta U = 0, \quad -\Delta V = V^{2^*-1}, \quad 0 \leq U, V \leq 1 = \max\{U(0), V(0)\}.$$

If $U(0) = 1$, then by Liouville's theorem we have $U(x) \equiv 1$. However,

$$\int\limits_{\mathbb{R}^N} U^{2^*} \, dx \leq \lim_{n \to +\infty} \int\limits_{\mathbb{R}^N} U_n^{2^*} \, dx = \lim_{n \to +\infty} K_n^{-N\varepsilon_n/2} \int\limits_{\mathbb{R}^N} u_n^{2^*} \, dx < +\infty,$$

which is a contradiction. So $V(0) = 1$, and $V \in D^{1,2}(\mathbb{R}^N)$ is a positive solution of $-\Delta v = v^{2^*-1}$, $v \in D^{1,2}(\mathbb{R}^N)$. This implies $|V|_{2^*}^{2^*} = S^{N/2}$ and so

$$\frac{1}{N} S^{N/2} = \frac{1}{N} \int\limits_{\mathbb{R}^N} |V|^{2^*} \, dx \leq \limsup_{n \to +\infty} \left(\frac{1}{2} - \frac{1}{2^* - \varepsilon_n} \right) K_n^{\frac{N-2}{2}\varepsilon_n} \int\limits_{\mathbb{R}^N} |V_n|^{2^*-\varepsilon_n} \, dx$$

$$= \limsup_{n \to +\infty} \left(\frac{1}{2} - \frac{1}{2^* - \varepsilon_n} \right) \int\limits_{\mathbb{R}^N} |v_n|^{2^*-\varepsilon_n} \, dx$$

$$\leq \limsup_{n \to +\infty} \left[\left(\frac{1}{2} - \frac{1}{2^* - \varepsilon_n} \right) |v_n|_{2^*-\varepsilon_n}^{2^*-\varepsilon_n} + \left(\frac{1}{2} - \frac{1}{p+1} \right) |u_n|_{p+1}^{p+1} \right]$$

$$= \limsup_{n \to +\infty} c_n \leq m < \frac{1}{N} S^{N/2},$$

which is also a contradiction. Therefore, $\{u_n, v_n\}_{n \in \mathbb{N}}$ is bounded in $L^\infty(\mathbb{R}^N)$. This implies that

$$\lim_{n \to +\infty} \int\limits_{\mathbb{R}^N} |v_n|^{2^*-\varepsilon_n} \, dx = \int\limits_{\mathbb{R}^N} |v_0|^{2^*} \, dx. \tag{6.29}$$

In fact, since $(v_n)_{n \geq 0}$ are radial and bounded in $H^1(\mathbb{R}^N)$, we see that

$$v_n(x) \to 0 \quad \text{as } |x| \to +\infty \quad \text{uniformly for } n \in \{0\} \cup \mathbb{N}.$$

Therefore, for any $\varepsilon > 0$, there exists $R_\varepsilon > 0$ such that for any $n \in \{0\} \cup \mathbb{N}$ there holds (let $\varepsilon_0 = 0$)

$$\int_{\mathbb{R}^N \setminus B(0, R_\varepsilon)} |v_n|^{2^* - \varepsilon_n} \, dx \leq \varepsilon \int_{\mathbb{R}^N \setminus B(0, R_\varepsilon)} |v_n|^2 \, dx \leq C\varepsilon,$$

where $B(0, R) := \{x \in \mathbb{R}^N : |x| < R\}$. On the other hand, since $\{v_n\}_{n \in \mathbb{N}}$ is bounded in $L^\infty(\mathbb{R}^N)$, the Lebesgue's dominated convergent theorem implies

$$\lim_{n \to +\infty} \int_{B(0, R_\varepsilon)} |v_n|^{2^* - \varepsilon_n} \, dx = \int_{B(0, R_\varepsilon)} |v_0|^{2^*} \, dx.$$

So (6.29) holds. On the other hand, $J_n'(u_n, v_n)(u_n, v_n) = 0$ implies that $\{(u_n, v_n)\}_{n \in \mathbb{N}}$ is bounded away from $(0, 0)$ in H. Combining this with (6.27) we see that $\inf\limits_{n \in \mathbb{N}} c_n > 0$.
Since (u_0, v_0) satisfies (6.4), we have

$$\begin{aligned}
I(u_0, v_0) &= \left(\frac{1}{2} - \frac{1}{2^*}\right) \int_{\mathbb{R}^N} |v_0|^{2^*} \, dx + \left(\frac{1}{2} - \frac{1}{p+1}\right) \int_{\mathbb{R}^N} |u_0|^{p+1} \, dx \\
&= \lim_{n \to +\infty} \left[\left(\frac{1}{2} - \frac{1}{2^* - \varepsilon_n}\right) |v_n|_{2^* - \varepsilon_n}^{2^* - \varepsilon_n} + \left(\frac{1}{2} - \frac{1}{p+1}\right) |u_n|_{p+1}^{p+1} \right] \\
&= \lim_{n \to +\infty} c_n \in (0, m].
\end{aligned}$$

Therefore, $(u_0, v_0) \neq (0, 0)$ and $(u_0, v_0) \in M$, which implies that $I(u_0, v_0) = m = m_{\mu,\nu,\lambda}$, namely (u_0, v_0) is a ground state solution of (6.4), and $u_0 \neq 0$, $v_0 \neq 0$. By (6.28) and (6.29) we have

$$\|u_0\|_\mu^2 + \|v_0\|_\nu^2 - 2\lambda \int_{\mathbb{R}^N} u_0 v_0 \, dx = \lim_{n \to +\infty} \left(\|u_n\|_\mu^2 + \|v_n\|_\nu^2 - 2\lambda \int_{\mathbb{R}^N} u_n v_n \, dx \right),$$

which implies that $(u_n, v_n) \to (u_0, v_0)$ strongly in H. Since $\{u_n, v_n\}_{n \in \mathbb{N}}$ is bounded in $L^\infty(\mathbb{R}^N)$, we have $u_0, v_0 \in L^\infty(\mathbb{R}^N)$. Then by the standard elliptic regularity theory, $u_0, v_0 \in C^2(\mathbb{R}^N, \mathbb{R})$. Since u_n, v_n are positive radial symmetric decreasing, we see that $u_0 \geq 0$, $v_0 \geq 0$ are radial symmetric nonincreasing. Finally, the strong maximum principle yields that u_0, v_0 are positive radial symmetric decreasing. $\square$

Proof (Proof of Theorem 6.1) Theorem 6.1 follows directly from Lemmas 6.4, 6.5 and 6.6.   □

Remark 6.4 In the case $\mu > \mu_0$, we see from Lemma 6.4 that $m_{\mu,\nu,\lambda_{\mu,\nu}} = \frac{1}{N}S^{N/2}$ and $m_{\mu,\nu,\lambda} < \frac{1}{N}S^{N/2}$ for $\lambda > \lambda_{\mu,\nu}$, and so the methods of proving Lemma 6.5 and Lemma 6.6 cannot be used in case $\lambda = \lambda_{\mu,\nu}$. Hence, the existence of the ground state solutions for this case remains an open question.

6.3 Asymptotics

In this section, we give the proof of Theorem 6.2.

Proof (Proof of Theorem 6.2) Assume that μ, ν, λ satisfy the hypotheses in (1) or (2)-(ii) of Theorem 6.1, and let (u, v) be any a ground state solution of (6.4). Then $u \neq 0, v \neq 0$. Define $u^+ := \max\{u, 0\}$ and $u^- := \max\{-u, 0\}$. Without loss of generality, we may assume that $(u^+, v^+) \neq (0, 0)$. By (6.12) and (6.13) we see that $t_{\lambda,|u|,|v|} \leq t_{\lambda,u,v} = 1$, and

$$m := m_{\mu,\nu,\lambda} \leq I(t_{\lambda,|u|,|v|}|u|, t_{\lambda,|u|,|v|}|v|) \leq I(u, v) = m.$$

It follows that $t_{\lambda,|u|,|v|} = 1$, namely $\int_{\mathbb{R}^N} uv\mathrm{d}x = \int_{\mathbb{R}^N} |u||v|\,\mathrm{d}x$. Combining this with $I'(u, v)(u^+, v^+) = 0$ we get that $(u^+, v^+) \in M$. Then $I(u, v) \leq I(u^+, v^+)$, and so $(u^-, v^-) = (0, 0)$, namely both $u \geq 0$ and $v \geq 0$.

     Remark that system (6.4) is cooperative (the definition can be seen in [20]) since $0 < \lambda < \sqrt{\mu\nu}$. Then by Busca and Sirakov [20] we have that (u, v) is radial up to a translation. So we may assume that (u, v) is radial symmetric.

     Assume that there exists $(u_n, v_n)_{n\in\mathbb{N}}$ such that they are positive radial symmetric ground state solutions of (6.4) and $u_n(0) + v_n(0) = \max\limits_{x\in\mathbb{R}^N} u_n(x) + \max\limits_{x\in\mathbb{R}^N} v_n(x) \to +\infty$ as $n \to +\infty$. Then by a similar blowup analysis as in Lemma 6.6, we get a contradiction. Thus, there exists a positive constant $C = C(\mu, \nu, \lambda)$ independent of (u, v) such that

$$\|u\|_{L^\infty(\mathbb{R}^N)} + \|v\|_{L^\infty(\mathbb{R}^N)} \leq C.$$

     By the standard elliptic regularity theory, we have $u, v \in C^2(\mathbb{R}^N, \mathbb{R})$. Hence, by the strong maximum principle, we have $u, v > 0$.

     By a similar argument as [31, Lemma 4.5], we see that u, v are both decreasing with respect to $r = |x| \in [0, +\infty)$.

     Finally, fix any $\mu \in (0, \mu_0), \nu > 0$ and let $\lambda_n \in (0, \sqrt{\mu\nu})$ such that $\lambda_n \to 0$ as $n \to +\infty$. Let $(u_{\lambda_n}, v_{\lambda_n})$ be any positive radial ground state solution of (6.4) with $\lambda = \lambda_n$. By the proof of (1) in Lemma 6.4 we have $m_{\mu,\nu,\lambda_n} \leq f_\mu(w_\mu) < \frac{1}{N}S^{N/2}$. Similarly to (6.27), we see that $\{(u_{\lambda_n}, v_{\lambda_n})\}_{n\in\mathbb{N}}$ is bounded in H. Then passing to a subsequence, we may assume that $(u_{\lambda_n}, v_{\lambda_n}) \rightharpoonup (u_0, v_0)$ weakly in H, and so (u_0, v_0) satisfies

$$\begin{cases} -\Delta u + \mu u = |u|^{p-1}u, & x \in \mathbb{R}^N, \\ -\Delta v + vv = |v|^{2^*-2}v, & x \in \mathbb{R}^N, \\ u, v \in H^1(\mathbb{R}^N). \end{cases}$$

This means $v_0 \equiv 0$. Since u_{λ_n} is radial, one has that u_0 is radial and so

$$\lim_{n \to +\infty} \int_{\mathbb{R}^N} |u_{\lambda_n}|^{p+1} \, dx = \int_{\mathbb{R}^N} |u_0|^{p+1} \, dx.$$

We claim that $u_0 \not\equiv 0$. Assume by contradiction that $u_0 \equiv 0$. By the Pohozaev identity and $I'_{\mu,\nu,\lambda_n}(u_{\lambda_n}, v_{\lambda_n})(u_{\lambda_n}, v_{\lambda_n}) = 0$ we easily obtain

$$\left(\frac{2^*}{p+1} - 1 \right) |u_{\lambda_n}|_{p+1}^{p+1} = \frac{2^* - 2}{2} \left(\mu |u_{\lambda_n}|_2^2 + v |v_{\lambda_n}|_2^2 - 2\lambda_n \int_{\mathbb{R}^N} u_{\lambda_n} v_{\lambda_n} \, dx \right).$$

Letting $n \to +\infty$ we have $\mu |u_{\lambda_n}|_2^2 + v |v_{\lambda_n}|_2^2 \to 0$. By $I'_{\mu,\nu,\lambda_n}(u_{\lambda_n}, v_{\lambda_n})(0, v_{\lambda_n}) = 0$ we have

$$\lim_{n \to +\infty} \left(\int_{\mathbb{R}^N} |\nabla v_{\lambda_n}|^2 \, dx - |v_{\lambda_n}|_{2^*}^{2^*} \right) = 0.$$

By (6.30), $\{|v_{\lambda_n}|_{2^*}^{2^*}\}_{n \in \mathbb{N}}$ is bounded. Then, passing to a subsequence, we may assume that

$$\lim_{n \to +\infty} \int_{\mathbb{R}^N} |\nabla v_{\lambda_n}|^2 \, dx = \lim_{n \to +\infty} |v_{\lambda_n}|_{2^*}^{2^*} =: B_0.$$

From Lemma 6.1 we have $\liminf\limits_{n \to +\infty} m_{\mu,\nu,\lambda_n} \geq m_{\mu,\nu,\lambda_1} > 0$ and so $B_0 > 0$. Using Sobolev inequality (6.7), we have $B_0 \geq \frac{1}{N} S^{N/2}$. Using (6.30) again, we have $\frac{1}{N} B_0 \leq f_\mu(w_\mu) < \frac{1}{N} S^{N/2}$, a contradiction. Thus $u_0 \not\equiv 0$ and so

$$f_\mu(w_\mu) \leq f_\mu(u_0) = \left(\frac{1}{2} - \frac{1}{p+1} \right) |u_0|_{p+1}^{p+1}$$

$$\leq \limsup_{n \to +\infty} \left[\left(\frac{1}{2} - \frac{1}{2^*} \right) |v_{\lambda_n}|_{2^*}^{2^*} + \left(\frac{1}{2} - \frac{1}{p+1} \right) |u_{\lambda_n}|_{p+1}^{p+1} \right]$$

$$= \limsup_{n \to +\infty} m_{\mu,\nu,\lambda_n} \leq f_\mu(w_\mu). \tag{6.30}$$

This implies $f_\mu(u_0) = f_\mu(w_\mu)$ and so u_0 is a positive radial ground state solution of

$$-\Delta u + \mu u = |u|^{p-1}u, \quad u \in H^1(\mathbb{R}^N).$$

Moreover, we have $|v_{\lambda_n}|_{2^*}^{2^*} \to 0$ as $n \to +\infty$. Thus

$$\|u_0\|_\mu^2 = \limsup_{n \to +\infty} \left(\|u_{\lambda_n}\|_\mu^2 + \|v_{\lambda_n}\|_\nu^2 - 2\lambda_n \int_{\mathbb{R}^N} u_{\lambda_n} v_{\lambda_n} \, dx \right),$$

which implies $(u_{\lambda_n}, v_{\lambda_n}) \to (u_0, 0)$ strongly in H. This completes the proof. $\square$

References

1. Abdellaoui, B., Felli, V., Peral, I.: Some remarks on systems of elliptic equations doubly critical in the whole $\mathbb{R}^N$. Calc. Var. Partial Differ. Equ. **34**, 97–137 (2009)
2. Adimurthi. Yadava, S.: An elementary proof of the uniqueness of positive radial solutions of a quasilinear dirichlet problem. Arch. Ration. Mech. Anal. **127**, 219–229 (1994)
3. Akhmediev, N., Ankiewicz, A.: Partially coherent solitons on a finite background. Phys. Rev. Lett. **82**, 2611–2664 (1999)
4. Akhmediev, N., Ankiewicz, A.: Novel soliton states and bifurcation phenomena in nonlinear fiber couplers. Phys. Rev. Lett. **70**, 2395–2398 (1993)
5. Ambrosetti, A.: Remarks on some systems of nonlinear Schrödinger equations. Fixed Point Theory Appl. **4**, 35–46 (2008)
6. Ambrosetti, A., Colorado, E.: Bound and ground states of coupled nonlinear Schrödinger equations. C. R. Math. Acad. Sci. Paris **342**, 453–458 (2006)
7. Ambrosetti, A., Colorado, E.: Standing waves of some coupled nonlinear Schrödinger equations. J. Lond. Math. Soc. **75**, 67–82 (2007)
8. Ambrosetti, A., Colorado, E., Ruiz, D.: Multi-bump solitons to linearly coupled systems of nonlinear Schrödinger equations. Calc. Var. Partial Differ. Equ. **30**, 85–112 (2007)
9. Ambrosetti, A., Cerami, G., Ruiz, D.: Solitons of linearly coupled systems of semilinear non-autonomous equations on $\mathbb{R}^N$. J. Funct. Anal. **254**, 2816–2845 (2008)
10. Ambrosotti, A.; Rabinowitz, P.: Dual variational methods in critical point theory and applications. J. Funct. Anal. **14**, 349–381 (1973)
11. Atkinson, F., Peletier, L.: Emden-Fowler equations involving critical exponents. Nonlinear Anal. **10**, 755–776 (1986)
12. Aubin, T.: Problemes isoperimetriques et espaces de Sobolev. J. Differ. Geom. **11**, 573–598 (1976)
13. Bartsch, T., Dancer, N., Wang, Z.: A Liouville theorem, a priori bounds, and bifurcating branches of positive solutions for a nonlinear elliptic system. Calc. Var. Partial Differ. Equ. **37**, 345–361 (2010)
14. Bartsch, T., Wang, Z.: Note on ground states of nonlinear Schrödinger systems. J. Partial Differ. Equ. **19**, 200–207 (2006)
15. Bartsch, T., Wang, Z., Wei, J.: Bound states for a coupled Schrödinger system. J. Fixed Point Theory Appl. **2**, 353–367 (2007)
16. Bartsch, T., Willem, M.: Infinitely many radial solutions of a semilinear elliptic problem on $\mathbb{R}^N$. Arch. Ration. Mech. Anal. **124**, 261–276 (1993)
17. Brezis, H., Kato, T.: Remarks on the Schrödinger operator with singular complex potentials. J. Math. Pures Appl. **58**, 137–151 (1979)

© Springer-Verlag Berlin Heidelberg 2015
Z. Chen, *Solutions of Nonlinear Schrödinger Systems*, Springer Theses,
DOI 10.1007/978-3-662-45478-7

18. Brezis, H., Lieb, E.: Minimum action solutions of some vector field equations. Comm. Math. Phys. **96**, 97–113 (1984)
19. Brezis, H., Nirenberg, L.: Positive solutions of nonlinear elliptic equations involving critical Sobolev exponents. Comm. Pure Appl. Math. **36**, 437–477 (1983)
20. Busca, J., Sirakov, B.: Symmetry results for semilinear elliptic systems in the whole space. J. Differ. Equ. **163**, 41–56 (2000).
21. Byeon, J., Jeanjean, L., Maris, M.: Symmetry and monotonicity of least energy solutions. Calc. Var. Partial Differ. Equ. **36**, 481–492 (2009)
22. Castro, A., Cossio, J., Neuberger, J.: A sign-changing solution for a superlinear Dirichlet problem. Rocky Mountain J. Math. **27**, 1041–1053 (1997)
23. Chang, S., Lin, C.-S., Lin, T., Lin, W.: Segregated nodal domains of two-dimensional multi-species Bose-Einstein condensates. Phys. D. **196**, 341–361 (2004)
24. Cerami, G., Solimini, S., Struwe, M.: Some existence results for superlinear elliptic boundary value problems involving critical exponents. J. Funct. Anal. **69**, 289–306 (1986)
25. Chen, Z., Lin, C.-S.: Asymptotic behavior of least energy solutions for a critical elliptic system. Preprint, (2014)
26. Chen, Z., Lin, C.-S., Zou, W.: Infinitely many sign-changing and semi-nodal solutions for a nonlinear Schrödinger system. Ann. Scuola Norm. Sup. Pisa Cl. Sci. (2014). doi:10.2422/2036-2145.201401_002
27. Chen, Z., Lin, C.-S., Zou, W.: Multiple sign-changing and semi-nodal solutions for coupled Schrödinger equations. J. Differ. Equ. **255**, 4289–4311 (2013)
28. Chen, Z., Lin, C.-S., Zou, W.: Monotonicity and nonexistence results to cooperative systems in the half space. J. Funct. Anal. **266**, 1088–1105 (2014)
29. Chen, Z., Lin, C.-S., Zou, W.: Sign-changing solutions and phase separation for an elliptic system with critical exponent. Comm. Partial Differ. Equ. **39**, 1827–1859 (2014)
30. Chen, Z., Shioji, N., Zou, W.: Ground state and multiple solutions for a critical exponent problem. Nonlinear Differ. Equ. Appl. **19**, 253–277 (2012)
31. Chen, Z., Zou, W.: On coupled systems of Schrödinger equations. Adv. Differ. Equ. **16**, 775–800 (2011)
32. Chen, Z., Zou, W.: On the Brezis-Nirenberg problem in a ball. Differ. Integr. Equ. **25**, 527–542 (2012)
33. Chen, Z., Zou, W.: Positive least energy solutions and phase separation for coupled Schrödinger equations with critical exponent. Arch. Ration. Mech. Anal. **205**, 515–551 (2012)
34. Chen, Z., Zou, W.: Ground states for a system of Schrödinger equations with critical exponent. J. Funct. Anal. **262**, 3091–3107 (2012)
35. Chen, Z., Zou, W.: An optimal constant for the existence of least energy solutions of a coupled Schrödinger system. Calc. Var. Partial Differ. Equ. **48**, 695–711 (2013)
36. Chen, Z., Zou, W.: Standing waves for coupled nonlinear Schrödinger equations with decaying potentials. J. Math. Phys. **54**, 111505 (2013)
37. Chen, Z., Zou, W.: On linearly coupled Schrödinger systems. Proc. Am. Math. Soc. **142**, 323–333 (2014)
38. Chen, Z., Zou, W.: Standing waves for linearly coupled Schrödinger equations with critical exponent. Ann. Inst. Henri Poincaré Anal. Non Linéaire. **31**, 429–447 (2014)
39. Chen, Z., Zou, W.: Existence and symmetry of positive ground states for a doubly critical Schrödinger system. Trans. Am. Math. Soc. (2014). doi:10.1090/S0002-9947-2014-06237-5
40. Chen, Z., Zou, W.: Standing waves for a coupled system of nonlinear Schrödinger equations. Annali di Matematica (2013). doi:10.1007/s10231-013-0371-5
41. Chen, Z., Zou, W.: A remark on doubly critical elliptic systems. Calc. Var. Partial Differ. Equ. **50**, 939–965 (2014)
42. Chen, Z., Zou, W.: Positive least energy solutions and phase separation for coupled Schrödinger equations with critical exponent: higher dimensional case. Calc. Var. Partial Differ. Equ. (2014). doi:10.1007/s00526-014-0717-x
43. Clapp, M., Weth, T.: Multiple solutions for the Brezis-Nirenberg problem. Adv. Differ. Equ. **10**, 463–480 (2005)

44. Crandall, M., Rabinowitz, P.: Bifurcation form simple eigenvalues. J. Funct. Anal. **8**, 321–340 (1971)
45. Crandall, M., Rabinowitz, P.: Bifurcation, perturbation of simple eigenvalues and linearized stability. Arch. Ration. Mech. Anal. **52**, 161–180 (1973)
46. Dancer, E., Wang, K., Zhang, Z.: The limit equation for the Gross-Pitaevskii equations and S.Terracini's conjecture. J. Funct. Anal. **262**, 1087–1131 (2012)
47. Dancer, E., Wei, J.: Spike solutions in coupled nonlinear Schrödinger equations with attractive interaciton. Trans. Am. Math. Soc. **361**, 1189–1208 (2009)
48. Dancer, N., Wei, J., Weth, T.: A priori bounds versus multiple existence of positive solutions for a nonlinear Schrödinger system. Ann. Inst. Henri Poincaré Anal. Non Linéaire **27**, 953–969 (2010)
49. de Figueiredo, D., Lopes, O.: Solitary waves for some nonlinear Schrödinger systems. Ann. Inst. Henri Poincaré Anal. Non Linéaire **25**, 149–161 (2008)
50. Devillanova, G., Solimini, S.: Concentration estimates and multiple solutions to elliptic problems at critical growth. Adv. Differ. Equ. **7**, 1257–1280 (2002)
51. Esry, B., Greene, C., Burke, J., Bohn, J.: Hartree-Fock theory for double condesates. Phys. Rev. Lett. **78**, 3594–3597 (1997)
52. Frantzeskakis, D.: Dark solitons in atomic Bose-Einstein condesates: from theory to experiments. J. Phys. A. **43**, 213001 (2010)
53. Gidas, B., Ni, W., Nirenberg, L.: Symmetry and related properties via the maximum principle. Comm. Math. Phys. **68**, 209–243 (1979)
54. Gilbarg, D., Trudinger, N.: Elliptic partial differential equations of second order, 2nd edn. Springer, Berlin (1983)
55. Guo, Y., Wei, J.: Infinitely many positive solutions for nonlinear Schrödinger system with nonsymmetric first order (2013). http://www.math.cuhk.edu.hk/wei/publicationpreprint
56. Hirano, N.: Multiple existence of nonradial positive solutions for a coupled nonlinear Schrödinger system. Nonlinear Differ. Equ. Appl. **16**, 159–188 (2009)
57. Ikoma, N., Tanaka, K.: A local mountain pass type result for a system of nonlinear Schrödinger equations. Calc. Var. Partial Differ. Equ. **40**, 449–480 (2011)
58. Kim, S.: On vector solutions for coupled nonlinear Schrödinger equations with critical exponents. Comm. Pure Appl. Anal. **12**, 1259–1277 (2013)
59. Kivshar, Y., Luther-Davies, B.: Dark optical solitons: physics and applications. Phys. Rep. **298**, 81–197 (1998)
60. Kwong, M.: Uniqueness of positive solutions of $\Delta u - u + u^p = 0$ in $\mathbb{R}^N$. Arch. Ration. Mech. Anal. **105**, 243–266 (1989)
61. Lin, T., Wei, J.: Ground state of N coupled nonlinear Schrödinger equations in $\mathbb{R}^n$, $n \le 3$. Comm. Math. Phys. **255**, 629–653 (2005)
62. Lin, T., Wei, J.: Spikes in two coupled nonlinear Schrödinger equations. Ann. Inst. Henri Poincaré Anal. Non Linéaire **22**, 403–439 (2005).
63. Lin, T., Wei, J.: Spikes in two-component systems of nonlinear Schrödinger equations with trapping potentials. J. Differ. Equ. **229**, 538–569 (2006)
64. Liu, J., Liu, X., Wang, Z.: Multiple mixed states of nodal solutions for nonlinear Schrödinger systems. Calc. Var. Partial Differ. Equ. (2014). doi:10.1007/s00526-014-0724-y
65. Liu, Z., Wang, Z.: Multiple bound states of nonlinear Schrödinger systems. Comm. Math. Phys. **282**, 721–731 (2008)
66. Liu, Z., Wang, Z.: On the Ambrosetti-Rabinowitz superlinear condition. Adv. Nonlinear Stud. **4**, 563–574 (2004)
67. Maia, L., Montefusco, E., Pellacci, B.: Positive solutions for a weakly coupled nonlinear Schrödinger systems. J. Differ. Equ. **229**, 743–767 (2006)
68. Maia, L., Montefusco, E., Pellacci, B.: Infinitely many nodal solutions for a weakly coupled nonlinear Schrödinger system. Commun. Contemp. Math. **10**, 651–669 (2008)
69. Maia, L., Pellacci, B., Squassina, M.: Semiclassical states for weakly coupled nonlinear Schrödinger systems. J. Eur. Math. Soc. **10**, 47–71 (2007)

70. Mitchell, M., Chen, Z., Shih, M., Segev, M.: Self-trapping of partially spatially incoherent light. Phys. Rev. Lett. **77**, 490–493 (1996)
71. Mitchell, M., Segev, M.: Self-trapping of incoherent white light. Nature **387**, 880–882 (1997)
72. Noris, B., Ramos, M.: Existence and bounds of positive solutions for a nonlinear Schrödinger system. Proc. Am. Math. Soc. **138**, 1681–1692 (2010)
73. Noris, B., Tavares, H., Terracini, S., Verzini, G.: Uniform Hölder bounds for nonlinear Schrödinger systems with strong competition. Commun. Pure Appl. Math. **63**, 267–302 (2010)
74. Noris, B., Tavares, H., Terracini, S., Verzini, G.: Convergence of minimax and continuation of critical points for singularly perturbed systems. J. Eur. Math. Soc. **14**, 1245–1273 (2012)
75. Palais, R.: The principle of symmetric criticality. Comm. Math. Phys. **69**, 19–30 (1979)
76. Parkins, A., Walls, D.: The Physics of trapped dilute-gas Bose-Einstein condensates. Phys. Rep. **303**, 1–80 (1998)
77. Pomponio, A.: Coupled nonlinear Schrödinger systems with potentials. J. Differ. Equ. **226**, 258–281 (2006)
78. Sato, Y., Wang, Z.: On the multiple existence of semi-positive solutions for a nonlinear Schrödinger system. Ann. Inst. Henri Poincaré Anal. Non Linéaire **30**, 1–22 (2013)
79. Schechter, M., Zou, W.: On the Brezis-Nirenberg problem. Arch. Ration. Mech. Anal. **197**, 337–356 (2010)
80. Sirakov, B.: Least energy solitary waves for a system of nonlinear Schrödinger equations in $\mathbb{R}^n$. Commun. Math. Phys. **271**, 199–221 (2007)
81. Struwe, M.: Variational Methods-Applications to Nonlinear Partial Differential Equations and Hamiltonian Systems. Springer, Berlin (1996)
82. Talenti, G.: Best constant in Sobolev inequality. Ann. Mat. Pure Appl. **110**, 352–372 (1976)
83. Tavares, H., Terracini, S.: Sign-changing solutions of competition-diffusion elliptic systems and optimal partition problems. Ann. Inst. Henri Poincaré Anal. Non Linéaire **29**, 279–300 (2012)
84. Terracini, S., Verzini, G.: Multipulse phases in k-mixtures of Bose-Einstein condensates. Arch. Ration. Mech. Anal. **194**, 717–741 (2009)
85. Timmermans, E.: Phase separation of Bose-Einstein condensates. Phys. Rev. Lett. **81**, 5718–5721 (1998)
86. Wei, J., Weth, T.: Nonradial symmetric bound states for a system of two coupled Schrödinger equations. Rend. Lincei Mat. Appl. **18**, 279–293 (2007)
87. Wei, J., Weth, T.: Radial solutions and phase separation in a system of two coupled Schrödinger equations. Arch. Ration. Mech. Anal. **190**, 83–106 (2008)
88. Wei, J., Weth, T.: Asymptotic behaviour of solutions of planar elliptic systems with strong competition. Nonliearity **21**, 305–317 (2008)
89. Wei, J., Yao, W.: Uniqueness of positive solutions to some coupled nonlinear Schrödinger equations. Commun. Pure Appl. Anal. **11**, 1003–1011 (2012)
90. Willem, M.: Minimax theorems. Birkhäuser, Basel (1996)

Index

© Springer-Verlag Berlin Heidelberg 2015
Z. Chen, *Solutions of Nonlinear Schrödinger Systems*, Springer Theses,
DOI 10.1007/978-3-662-45478-7

Printed in the United States
By Bookmasters